The Ecology
of Human Development

The Ecology
of Human Development

EXPERIMENTS BY NATURE AND DESIGN

Urie Bronfenbrenner

HARVARD UNIVERSITY PRESS

Cambridge, Massachusetts, and London, England   1979

Library of Congress Cataloging in Publication Data

Bronfenbrenner, Urie, 1917–
    The ecology of human development.

    Bibliography: p.
    Includes index.
    1. Child psychology—Research. I. Title.
[DNLM: 1. Human development. BF713 B869e]
BF722.B76   155.4   78–27232
ISBN 0–674–22456–6

To those who taught me most—
my parents,
my wife,
my children,
and my grandchildren

Foreword

Goethe, who commented wisely on so many aspects of human experience, said of our attempts to understand the world

Everything has been thought of before,
The difficulty is to think of it again.

To this I would add (supposing that Goethe also said something to this effect, but not having discovered his discovery) that ideas are only as important as what you can do with them. Democrites supposed that the world was made up of atomic particles. Aside from his error in overlooking the implications of assuming that all atoms move in the same direction at the same rate, his astute guess about the atomic structure of matter did not have the same impact as Rutherford's rediscovery (with cloud chamber in hand) in 1900. In short, an idea is as powerful as what you can do with it.

Approximately one hundred years ago a number of scholars began to think that it would be possible to understand human psychological processes by conducting experiments, modeled on the precision and explicit, quantitative, data-analytic techniques that had propelled the physical sciences to such prominence in human affairs. Wilhelm Wundt is usually given the credit for this idea, although the *science* of psychology was born almost simultaneously in universities located in Germany, London, Cambridge (Massachusetts), and Kazan (U.S.S.R.).

What has been lost in our textbook accounts of the history of psychology is the fact that a great many other scholars who were around when psychology embraced the laboratory were not especially moved by the new enterprise. We tend to forget that Wundt himself believed that many psychological mysteries were beyond the reach of experimental methods, a belief not always shared by his more zealous followers. Even before dissension began to appear in the ranks of those who followed in Wundt's path, more serious reservations were voiced about the utility of laboratory techniques for explaining our inner workings. Wilhelm Dilthey was an early and eloquent critic of Wundt's "new" psychology. Dilthey, after long deliberation, concluded that psychology should give up its quest for general laws of human psychological processes. Instead, he advocated that we strive for a *descriptive* psychology that would capture the unique complexity of the individual with all of its idiosyncrasies. Dilthey believed that by reducing the complexity of human nature to carefully measured reaction times or minutely detailed introspective reports, Wundt and his followers accomplished little more than the interment of human psychological processes in a crypt fashioned of brass instruments.

Dilthey's position has not prevailed in academic psychology, and for good reason. His very enticing view of adequate psychological description has never satisfied us as a model for complete psychological analysis. The infinite tangles of past experience and present circumstances that make us what we are smother us in particulars, defying explanation or generalization; faced with such complexity, any plausible simplifying procedure can appear to be a lifeline.

Recognizing psychology's limitations, we joke that Henry James was the great psychologist, his brother William the novelist. Lamenting psychology's limitations, we nonetheless expect a proper scientific discipline to provide us with more systematic information about ourselves than a novel can. Lacking such a rigorous discipline, we have followed Wundt's narrower path in our methods, but the limitations of theory imposed by that choice do not rest easy. We are faced with the paradox of a successful science that tells us precious little about the concerns that beckon us to it. Those who engage in psychology as professionals either come to terms with its limitations or become bored with neat experiments, the significance of which remains too often obscure. Finding no promising alternatives, many choose inaction.

Although there have been many changes in the particulars of psychological theory since the time of Wundt and Dilthey, the two extreme approaches that generated the schism between descriptive and explanatory psychology in the first place have prevailed, as have their differences in sophistication of methods and acceptance as disciplines. Wundt's structuralism gave way to new schools of scientific psychology, each complete with its own structured, systematic, and constrained models and methods: Gestalt psychology, functionalism, behaviorism, and (most recently) experimental, cognitive psychology. Dilthey's criticism of this continuing effort to build a "nomothetic" psychological science has been rediscovered repeatedly, most recently in the humanistic psychologies of the late 1960s and 1970s, but each time without the crucial analytic tools for descriptive analyses or the power to explain what it describes.

Some few among psychology's practitioners, even in the earliest days, sought ways to link the descriptive and explanatory approaches, recognizing in this schism the seeds of psychology's undoing as a discipline. For example, in the early decades of the twentieth century it was common, especially in Germany, which gave birth to both movements, to encounter discussions of the "crisis" in psychology, for which various authors proposed various solutions. Coming on the heels of a decade of social and scientific activism in the 1960s (in which he took an active part) Urie Bronfenbrenner's work represents the continuation of efforts by this small, heterogeneous, but significant group of psychologists to overcome the "crisis" in psychology by constructing a discipline that is *both* experimental *and* descriptive of our lives as we know them.

His themes are those which concern everyone who hopes that psychology will shed light on our experience. The promise he offers us is very enticing. Psychology need not choose between rigor and relevance. It can do more than explain "strange behaviors in strange places." If properly pursued, it can tell us how those strange places and strange behaviors relate to the mundane contexts we refer to as our "everyday lives."

Professor Bronfenbrenner urges upon us his concern with specifying what people do in a way that will generalize beyond the contexts of our observations. He emphasizes the crucial importance of studying the environments within which we behave if we are ever to break away from particularistic descriptions and contentless processes. In both these concerns, he follows in the footsteps of very able predecessors.

But what should lead us to believe that Bronfenbrenner's prescriptions will succeed when the work of men whose ideas he has built on (Kurt Lewin, for example) seems to have disappeared—sunk into the sands of time or so absorbed into our collective folk wisdom that it is no longer extractable for purposes of analysis? The answer lies in his specification of procedures that are enough like what we already do to make them comprehensible, yet different enough to provide a better approximation to real-life phenomena.

Almost everyone who has read about psychological experiments has had occasion to puzzle over their meaning. Are Stanford students sadists or craven cowards as their behavior in Zimbardo's prison experiments suggests? Are people slaves to authority who would willingly inflict harm on helpless fellows as the Milgram studies of compliance tell us? Are people really indifferent to strangers in distress? Can IQ tests possibly tell us about the value of day care?

To each of these and many other questions Bronfenbrenner gives us the only honest answer imaginable—the same answer his grandmother would have offered had he been able to discuss these questions with her—"*it all depends.*" In technical language, "it all depends" translates into the idea that the explanations for what we do (assuming we achieve serviceable descriptions) are to be found in interactions between characteristics of people and their environments, past and present. As Bronfenbrenner says, "the main effects are in the interaction." He would also follow Kurt Lewin in suggesting that if we want to change behavior, we have to change environments.

All of these commonsense suggestions entail a reorientation of the way we think about psychological processes, which must come to be treated as properties of systems, systems in which the individual is but one element. These ideas will succeed if Bronfenbrenner has (to paraphrase him) irked and goaded enough able scholars by his audacious assertions into trying to prove him wrong. Systematic challenges, even if they should disable his specific assertions, would constitute success. These are ideas worth having again and again until we are ready to exploit their power. When that day arrives, psychology will become a unified science of human behavior.

Michael Cole

University of California, San Diego

Preface

In writing a volume of this kind, one becomes keenly aware that science is indeed a community of scholars. We stand on the shoulders of giants, and mistake the broadened vision for our own. In this instance the giants are Kurt Lewin, George Herbert Mead, Sigmund Freud, William I. and Dorothy S. Thomas, Edward C. Tolman, Lev Vygotsky, Kurt Goldstein, Otto Rank, Jean Piaget, and Ronald A. Fisher. From these I learned mainly by reading. There are others who struggled to teach me, often against resistance. Chief among them were my first teachers in psychology, Frank S. Freeman, Robert M. Ogden, and Walter Fenno Dearborn. Lauriston Sharp introduced me to cultural anthropology, Robert Ulich to philosophy, and Harry C. Carver to mathematical statistics and experimental design.

But the seeds of the ecological conceptions developed here had been planted long before I entered college. It was my good fortune to have been brought up on the premises of a state institution for those who were then called "the feebleminded," where my father was a neuropathologist. Along with his medical degree, he had a Ph.D. in zoology, and he was a field naturalist at heart. The institution grounds offered a rich biological and social terrain for his observant eye. There were over three thousand acres of farmland, wooded hills, moss-covered forest, and fetid swamp—all teeming with plant and animal life. In those days the institution was a functioning community; the patients spent most of their time out of the wards, not just in school classrooms but working on the farm and in the shops. There were cow, horse, pig, sheep, and chicken barns, a smithy, carpenter shops, a bakery, and a store house from which food and goods were delivered around the village in horse-drawn farm wagons driven by the inmates. All these activities are gone now—struck down by the courts as involuntary servitude.

That was the world of my childhood. My father took me on innumerable walks, from his laboratory through the wards, shops, and farmland—where he preferred to see and talk with his patients—and even more often beyond the barbed wire fence into the woods and hills that began at our doorstep. Wherever we were he would alert my unobservant eyes to the workings of nature by pointing to the functional interdependence between living organisms and their surroundings.

I remember especially vividly his anguish when the New York City courts would commit to our institution, out of error or—more probably—sheer desperation, perfectly normal children. Before he could unwind the necessary red tape to have them released, it would be too late. After a few weeks as one of eighty inmates in a cottage with two matrons, their scores on the intelligence tests administered as a compulsory part of the discharge process proved them mentally deficient: that meant remaining in the institution for the rest of their lives. There was a way out for these children, but the opportunity did not arise until they were much older. One of the places to which adult female inmates would be assigned to work was in the homes of staff, where they helped with housework, cooking, and child care. In this way, Hilda, Anna, and others after them became de facto members of our family and significant figures in my upbringing. But they seldom stayed for long. Just at the point when as a result of my mother's training in homemaking and their own everyday initiative they had become indispensable, my father would arrange for their discharge, for they could now pass the critical minimum on the all-determining Stanford-Binet.

It was a long time, however, before these concrete experiences were reflected in conscious ideas about an ecology of human development. These first began to emerge in an informal but intensive year-long weekly faculty seminar conducted thirty years ago. Ambitiously, my colleagues and I had sought to chart new horizons for theory and research in human development. The group included, among others, Robert B. MacLeod, Alexander Leighton, and Robin Williams. It was they who shook the intellectual foundations of a young investigator wedded to belief in the rigor of the laboratory and of psychometric methods. They opened my eyes to the power both of phenomenology and of social context. My knowledge of the latter was broadened in the course of three decades of collaborative research with my colleague Edward C. Devereux. To the two Charles R. Hendersons, father and son, I owe a continuing debt for lessons in the elegance and ecological adaptability of Fisherian designs.

Two sets of experiences gave form and substance to the new perspectives I had acquired in the faculty seminar. The first involved conducting field research in a cultural context. At first it had little impact on me, for with unconscious self-protectiveness I had chosen to work in familiar social terrain—a small rural community in upstate New York. Then my fellow seminar member Alexander Leighton persuaded me to join him for a summer as he began his now classic studies of community factors affecting mental health. It was under his tutelage, on the French coast of Nova Scotia, that I began a career of cross-cultural research in Western and Eastern Europe, the U.S.S.R., Israel, and elsewhere, including a mind-shattering glimpse of the People's Republic of China.

Experience in these societies had two profound effects on me that are reflected in the present volume. First, it radically expanded my awareness of the resilience, versatility, and promise of the species Homo sapiens as evidenced by its capacity to adapt to, tolerate, and especially create the ecologies in which it lives and grows. Seen in different contexts, human nature, which I had previously thought of as a singular noun, became plural and pluralistic; for the different environments were producing discernible differences, not only across but also within societies, in talent, temperament, human relations, and particularly in the ways in which the culture, or subculture, brought up its next generation. The process and product of making human beings human clearly varied by place and time. Viewed in historical as well as cross-cultural perspective, this diversity suggested the possibility of ecologies as yet untried that held a potential for human natures yet unseen, perhaps possessed of a wiser blend of power and compassion than has thus far been manifested.

Although this last prospect may appear a product of airy idealism, it is rooted in the harder ground of cross-cultural reality.

The second lesson I learned from work in other societies is that public policy has the power to affect the well-being and development of human beings by determining the conditions of their lives. This realization led to my heavy involvement during the past fifteen years in efforts to change, develop, and implement policies in my own country that could influence the lives of children and families. Participating in the Head Start Planning Committee, two Presidential Task Forces, and other scientific advisory groups at the national, state, and local levels, as well as testifying for and collaborating with politicians and government officials on legislation, brought me to an unexpected conclusion that is a recurrent theme in the pages that follow: concern with public policy on the part of researchers is essential for progress in the scientific study of human development.

These evolving ideas, whatever their merit, are no more the product of my own endeavors than of the patient and persistent efforts of my colleagues to open my eyes to the realities of the world in which they lived and worked. In the area of cross-cultural investigation, the following were among the most patient and persistent: Gerold Becker, Lydia Bozhovich, Zvi El-Peleg and family, Hsieh Ch'i-kang, Sophie Kav-Venaki, Kurt Lüscher and family, Richard and Gertrude Meili, Janusz Reykowski, Ruth Sharabany, Ron Shouval and family, Sandor Komlosi and family, Igor Kon, Aleksei Leontiev, Hartmut von Hentig, and Aleksander V. Zaporozhets.

In the interface between developmental research and public policy, my principal associates and mentors have been Birch Bayh, Orville G. Brim, John Brademas, Robert Cooke, David Goslin, Nicolas Hobbs, Sidney Johnson, Alfred Kahn, Mary Keyserling, Walter F. Mondale, Evelyn Moore, Albert Quie, Julius Richmond, John Scales, Sargent and Eunice Shriver, Jule Sugarman, Harold Watts, Sheldon White, and Edward Zigler.

This volume developed as part of a scholarly enterprise that I initiated several years ago with the counsel of a number of likeminded colleagues and with the material support of the Foundation of Child Development. Known as the Program on the Ecology of Human Development, the effort was undertaken with the aim of furthering theory, advanced training, and research in the actual environments in which human beings live and grow. Work on the book began while I was a Belding Scholar of the foundation.

In particular I express deep appreciation to Orville G. Brim, president of the Foundation of Child Development, and to Heidi Sigal, program associate, for their encouragement, wise advice, and active help in all aspects of the EHD Program, including the conception and preparation of this volume. In addition, a great debt, both intellectual and personal, is owed to the consultants to the program—Sarane Boocock, Michael Cole, Glen Elder, William Kessen, Melvin Kohn, Eleanor Maccoby, and Sheldon White. In countless letters, conversations, and phone calls, they communicated reactions and ideas that I have gradually assimilated as my own. I apologize to the extent that I have unwittingly failed to give them credit or—worse yet—to do justice to their thoughts.

I have also been fortunate in the generosity of numerous colleagues and students at Cornell and elsewhere who have been willing to read and criticize drafts of portions of the manuscript. They include Henry Alker, Irwin Altman, Jay Belsky, John Clausen, Moncrieff Cochran, Michael Cole, William Cross, Glen Elder, James Garbarino, Herbert Ginsburg, Stephen Hamilton, Melvin Kohn, Barbara Koslowski, Michael Lamb, Tom Lucas, Barbara Lust and her students, Kurt Lüscher, Eleanor Maccoby, Maureen Mahoney, Rudolf Moos, David Olds, Henry Ricciuti, Morris Stambler, Eric Wanner, John Weisz, Sheldon White, and one of the most astute critics, Liese Bronfenbrenner.

Two of these, Michael Cole and Eric Wanner, have also served as special and general editors of this volume. Their initiative, encouragement, and advice have improved the product and eased the perennial pains of an author's labor. Special appreciation is also expressed to the anonymous reviewers of separate chapters and total text, as well as to Harriet Moss for her thoughtful editing of the final manuscript. I also owe a scholar's debt to my friend and neighbor Geoffrey Bruun, who never forgets the source, or substance, of a quotation.

My sense of obligation and gratitude extends beyond individuals. It is a major thesis of this book that human abilities and their realization depend in significant degree on the larger social and institutional context of individual activity. This principle is especially applicable in the present instance. Since its founding, Cornell University, as a Land Grant Institution supported half by endowment and half by the state, has nurtured a tradition of freedom and responsibility and encouraged its faculty in moving beyond traditional disciplines to recognize that social science, if it is to achieve its own objectives, must be responsive to human needs and aspirations. This dual theme has found even fuller expression in the work of the New York State College of Human Ecology at Cornell under the creative leadership of three successive deans—David C. Knapp, Jean Failing, and Jerome Ziegler.

Finally, the greatest debt of all is to Joyce Brainard, who with the dedicated assistance of Mary Alexander, Stephen Kaufman, Mary Miller, and Kay Riddell supervised and carried out endless revisions of the manuscript with care, craftsmanship, and devotion.

The summary of Ogbu's research, included in chapter 10, was prepared by Stephen Hamilton for a jointly authored conference paper. Passages appearing in several chapters represent revisions of material previously published in *Child Development*, the *American Psychologist*, the *Journal of Social Issues*, and the *Zeitschrift für Soziologie*.

Urie Bronfenbrenner

Ithaca, New York

Contents

PART ONE

An Ecological Orientation

1.

Purpose and Perspective

In this volume, I offer a new theoretical perspective for research in human development. The perspective is new in its conception of the developing person, of the environment, and especially of the evolving interaction between the two. Thus development is defined in this work as a lasting change in the way in which a person perceives and deals with his environment. For this reason, it is necessary at the outset to give an indication of the somewhat unorthodox concept of the environment presented in this volume. Rather than begin with a formal exposition, I shall first introduce this concept by some concrete examples.

The ecological environment is conceived as a set of nested structures, each inside the next, like a set of Russian dolls. At the innermost level is the immediate setting containing the developing person. This can be the home, the classroom, or as often happens for research purposes—the laboratory or the testing room. So far we appear to be on familiar ground (although there is more to see than has thus far met the investigator's eye). The next step, however, already leads us off the beaten track for it requires looking beyond single settings to the relations between them. I shall argue that such interconnections can be as decisive for development as events taking place within a given setting. A child's ability to learn to read in the primary grades may depend no less on how he is taught than on the existence and nature of ties between the school and the home.

The third level of the ecological environment takes us yet farther afield and evokes a hypothesis that the person's development is profoundly affected by events occurring in settings in which the person is not even present. I shall examine data suggesting that among the most powerful influences affecting the development of young children in modern industrialized societies are the conditions of parental employment.

Finally, there is a striking phenomenon pertaining to settings at all three levels of the ecological environment outlined above: within any culture or subculture, settings of a given kind—such as homes, streets, or offices—tend to be very much alike, whereas between cultures they are distinctly different. It is as if within each society or subculture there existed a blueprint for the organization of every type of setting. Furthermore, the blueprint can be changed, with the result that the structure of the settings in a society can become markedly altered and produce corresponding changes in behavior and development. For example, research results suggest that a change in maternity ward practices affecting the relation between mother and newborn can produce effects still detectable five years later. In another case, a severe economic crisis occurring in a society is seen to have positive or negative impact on the subsequent development of children throughout the life span, depending on the age of the child at the time that the family suffered financial duress.

The detection of such wide-ranging developmental influences becomes possible only if one employs a theoretical model that permits them to be observed. Moreover, because such findings can have important implications both for science and for public policy, it is especially important that the theoretical model be methodologically rigorous, providing checks for validity and permitting the emergence of results contrary to the investigator's original hypotheses. The present volume represents an attempt to define the basic parameters of a theoretical model that meets these substantive and methodological requirements. The work also seeks to demonstrate the scientific utility of the ecological model for illuminating the findings of previous studies and for formulating new research problems and designs.

The environment as conceived in the proposed schema differs from earlier formulations not only in scope but also in content and structure. On the first count, the ecological orientation takes seriously and translates into operational terms a theoretical position often lauded in the literature of social science but seldom put into practice in research. This is the thesis, expounded by psychologists and sociologists alike, that what matters for behavior and development is the environment as it is *perceived* rather than as it may exist in "objective" reality. In the pages that follow, this principle is applied to expose both the weaknesses and the strengths of the laboratory and the testing room as contexts for assessing developmental processes. Evidence exists of consistent differences in the behavior of children and adults observed in the laboratory and in the actual settings of life. These differences in turn illuminate the various meanings of these types of settings to the participants, as partly a function of their social background and experience.

Different kinds of settings are also analyzed in terms of their structure. Here the approach departs in yet another respect from that of conventional research models: environments are not distinguished by reference to linear variables but are analyzed in systems terms. Beginning at the innermost level of the ecological schema, one of the basic units of analysis is the *dyad*, or two-person system. Although the literature of developmental psychology makes frequent reference to dyads as structures characterized by reciprocal relations, we shall see that, in practice, this principle is often disregarded. In keeping with the traditional focus of the laboratory procedure on a single experimental subject, data are typically collected about only one person at a time, for instance, about either the mother or the child but rarely for both simultaneously. In the few instances in which the latter does occur, the emerging picture reveals new and more dynamic possibilities for both parties. For instance, from dyadic data it appears that if one member of the pair undergoes a process of development, the other does also. Recognition of this relationship provides a key to understanding developmental changes not only in children but also in adults who serve as primary caregivers—mothers, fathers, grandparents, teachers, and so on. The same consideration applies to dyads involving husband and wife, brother and sister, boss and employee, friends, or fellow workers.

In addition, a systems model of the immediate situation extends beyond the dyad and accords equal developmental importance to what are called $N + 2$ *systems*—triads, tetrads, and larger interpersonal structures. Several findings indicate that the capacity of a dyad to serve as an effective context for human development is crucially dependent on the presence and participation of third parties, such as spouses, relatives, friends, and neighbors. If such third parties are absent, or if they play a disruptive rather than a supportive role, the developmental process, considered as a system, breaks down; like a three-legged stool, it is more easily upset if one leg is broken, or shorter than the others.

The same triadic principle applies to relations between settings. Thus the capacity of a setting—such as the home, school, or workplace—to function effectively as a context for development is seen to depend on the existence and nature of social interconnections between settings, including joint participation, communication, and the existence of information in each setting about the other. This principle accords importance to questions like the following: does a young person enter a new situation such as school, camp, or college alone, or in the company of familiar peers or adults? Are the person and her family provided with any information about or experience in the new setting before actual entry is made? How does such prior knowledge affect the subsequent course of behavior and development in the new setting?

Questions like these highlight the developmental significance and untapped research potential of what are called *ecological transitions*—shifts in role or setting, which occur throughout the life span. Examples of ecological transitions include the arrival of a younger sibling, entry into preschool or school, being promoted, graduating, finding a job, marrying, having a child, changing jobs, moving, and retiring.

The developmental importance of ecological transitions derives from the fact that they almost invariably involve a change in *role*, that is, in the expectations for behavior associated with particular positions in society. Roles have a magiclike power to alter how a person is treated, how she acts, what she does, and thereby even what she thinks and feels. The principle applies not only to the developing person but to the others in her world.

The environmental events that are the most immediate and potent in affecting a person's development are activities that are engaged in by others with that person or in her presence. Active engagement in, or even mere exposure to, what others are doing often inspires the person to undertake similar activities on her own. A three-year-old is more likely to learn to talk if others around her are talking and especially if they speak to her directly. Once the child herself begins to talk, it constitutes evidence that development has actually taken place in the form of a newly acquired *molar activity* (as opposed to molecular behavior, which is momentary and typically devoid of meaning or intent). Finally, the molar activities engaged in by a person constitute both the internal mechanisms and the external manifestations of psychological growth.

The sequence of nested ecological structures and their developmental significance can be illustrated with reference to the same example. We can hypothesize that a child is more likely to learn to talk in a setting containing roles that obligate adults to talk to children or that encourage or enable other persons to do so (such as when one parent does the chores so that the other can read the child a story).

But whether parents can perform effectively in their child-rearing roles within the family depends on role demands, stresses, and supports emanating from other settings. As we shall see, parents' evaluations of their own capacity to function, as well as their view of their child, are related to such external factors as flexibility of job schedules, adequacy of child care arrangements, the presence of friends and neighbors who can help out in large and small emergencies, the quality of health and social services, and neighborhood safety. The availability of supportive settings is, in turn, a function of their existence and frequency in a given culture or subculture. This frequency can be enhanced by the adoption of public policies and practices that create additional settings and societal roles conducive to family life.

A theoretical conception of the environment extending beyond the behavior of individuals to encompass functional systems both within and between settings, systems that can also be modified and expanded, contrasts sharply with prevailing research models. These established models typically employ a scientific lens that restricts, darkens, and even blinds the researcher's vision of environmental obstacles and opportunities and of the remarkable potential of human beings to respond constructively to an ecologically compatible milieu once it is made available. As a result, human capacities and strengths tend to be underestimated.

The structure of the ecological environment may also be defined in more abstract terms. As we have seen, the ecological environment is conceived as extending far beyond the immediate situation directly affecting the developing person—the objects to which he responds or the people with whom he interacts on a face-to-face basis. Regarded as of equal importance are connections between other persons present in the setting, the nature of these links, and their indirect influence on the developing person through their effect on those who deal with him at first hand. This complex of interrelations within the immediate setting is referred to as the *microsystem.*

The principle of interconnectedness is seen as applying not only within settings but with equal force and consequence to linkages between settings, both those in which the developing person actually participates and those that he may never enter but in which events occur that affect what happens in the person's immediate environment. The former constitute what I shall call *mesosystems,* and the latter *exosystems.*

Finally, the complex of nested, interconnected systems is viewed as a manifestation of overarching patterns of ideology and organization of the social institutions common to a particular culture or subculture. Such generalized patterns are referred to as *macrosystems.* Thus within a given society or social group, the structure and substance of micro-, meso-, and exosystems tend to be similar, as if they were constructed from the same master model, and the systems function in similar ways. Conversely, between different social groups, the constituent systems may vary markedly. Hence by analyzing and comparing the micro-, meso-, and exosystems characterizing different social classes, ethnic and religious groups, or entire societies, it becomes possible to describe systematically and to distinguish the ecological properties of these larger social contexts as environments for human development.

Most of the building blocks in the environmental aspect of the theory are familiar concepts in the behavioral and social sciences: molar activity, dyad, role, setting, social network, institution, subculture, culture. What is new is the way in which these entities are related to each other and to the course of development. In short, as far as the external world is concerned, what is presented here is a theory of environmental interconnections and their impact on the forces directly affecting psychological growth.

Furthermore, an ecological approach to the study of human development requires a reorientation of the conventional view of the proper relation between science and public policy. The traditional position, at least among social scientists, is that whenever possible social policy should be based on scientific knowledge. The line of thought I develop in this volume leads to a contrary thesis: in the interests of advancing fundamental research on human development, *basic science needs public policy even more than public policy needs basic science.* Moreover, what is required is not merely a complementary relation between these two domains but their functional integration. Knowledge and analysis of social policy are essential for progress in developmental research because they alert the investigator to those aspects of the environment, both immediate and more remote, that are most critical for the cognitive, emotional, and social development of the person. Such knowledge and analysis can also lay bare ideological assumptions underlying, and sometimes profoundly limiting, the formulation of research problems and designs and thus the range of possible findings. A functional integration between science and social policy of course does not mean that the two should be confused. In examining the impact of public policy issues for basic research in human development, it is all the more essential to distinguish between interpretations founded on empirical evidence and those rooted in ideological preference.

It is clear that the desirability of a reciprocal relation between science and social policy follows from the inclusion, in the theoretical model of the environment, of a macrosystem level involving generalized patterns of ideology and institutional structure characteristic of a particular culture or subculture. Public policy is a part of the macrosystem determining the specific properties of exo-, meso-, and microsystems that occur at the level of everyday life and steer the course of behavior and development.

Especially in its formal aspects, the conception of the environment as a set of regions each contained within the next draws heavily on the theories of Kurt Lewin (1935, 1936, 1941, 1948). Indeed this work may be viewed as an attempt to provide psychological and sociological substance to Lewin's brilliantly conceived topological territories.

Perhaps the most unorthodox feature of the proposed theory is its conception of development. Here the emphasis is not on the traditional psychological processes of perception, motivation, thinking, and learning, but on their *content—what* is perceived, desired, feared, thought about, or acquired as knowledge, and how the nature of this psychological material changes as a function of a person's exposure to and interaction with the environment. Development is defined as the person's evolving conception of the ecological environment, and his relation to it, as well as the person's growing capacity to discover, sustain, or alter its properties. Once again, this formulation shows the influence of Lewin, especially of his emphasis on a close interconnection and isomorphism between the structure of the person and of the situation (1935). The proposed conception also leans heavily on the ideas of Piaget, particularly as set forth in *The construction of reality in the child* (1954). The present thesis, however, goes considerably further. By contrast with Piaget's essentially "decontextualized" organism, it emphasizes the evolving nature and scope of perceived reality as it emerges and expands in the child's awareness and in his active involvement with the physical and social environment. Thus the infant at first becomes conscious only of events in his immediate surroundings, in what I have called the microsystem. Within this proximal domain, the focus of attention and of developing activity tends initially to be limited even more narrowly to events, persons, and objects that directly impinge on the infant. Only later does the young child become aware of relations between events and persons in the setting that do not from the outset involve his active participation. In the beginning the infant is also conscious of only one setting at a time, the one that he occupies at the moment. My own treatment of development not only includes the infant's awareness of the continuity of persons across settings, as implied by Piaget's concept of perceptual constancy, but also encompasses his dawning realization of the relations between events in different settings. In this way the developing child begins to recognize the existence and to develop an emerging sense of the mesosystem. The recognition of the possibility of relations between settings, coupled with the capacity to understand spoken and written language, enables him to comprehend the occurrence and nature of events in settings that he has not yet entered himself, like school, or those that he may never enter at all, such as the parents' workplace, a location in a foreign land, or the world of someone else's fantasy as expressed in a story, play, or film.

As Piaget emphasized, the child also becomes capable of creating and imagining a world of his own that likewise reflects his psychological growth. Again, an ecological perspective accords to this fantasy world both a structure and a developmental trajectory, for the realm of the child's imagination also expands along a continuum from the micro- to the meso-, exo-, and even macro- level.

The development of the child's fantasy world underscores the fact that his emerging perceptions and activities are not merely a reflection of what he sees but have an active, creative aspect. To use Piaget's apt term, the child's evolving phenomenological world is truly a "construction of reality" rather than a mere representation of it. As both Lewin and Piaget point out, the young child at first confuses the subjective and objective features of the environment and as a result can experience frustration, or even bodily harm, as he attempts the physically impossible. But gradually he becomes capable of adapting his imagination to the constraints of objective reality and even of refashioning the environment so that it is more compatible with his abilities, needs, and desires. It is this growing capacity to remold reality in accordance with human requirements and aspirations that, from an ecological perspective, represents the highest expression of development.

In terms of research method, the child's evolving construction of reality cannot be observed directly; it can only be inferred from patterns of activity as these are expressed in both verbal and non-verbal behavior, particularly in the activities, roles, and relations in which the person engages. These three factors also constitute what are designated as the *elements* of the microsystem.

In sum, this volume represents an attempt at theoretical integration. It seeks to provide a unified but highly differentiated conceptual scheme for describing and interrelating structures and processes in both the immediate and more remote environment as it shapes the course of human development throughout the life span. This integrative effort is regarded as the necessary first step in the systematic study of human development in its human context.

Throughout the volume, theoretical ideas are presented in the form of definitions of basic concepts, propositions which, in effect, constitute the axioms of the theory, and hypotheses that posit processes and relationships subject to empirical investigation.

Although some of the hypotheses to be proposed are purely deductive, following logically from defined concepts and stated propositions, the great majority derive from the application of the proposed theoretical framework to concrete empirical investigations. Thus I have by no means limited myself to theoretical exposition. I have made an effort throughout to translate ideas into operational terms. First, I have tried to find studies that illustrate the issues in question either by demonstration, or failing that, by default—by pointing out what the investigators might have done. Second, I have used investigations already published or reported to show in what way the results can be illuminated by applying concepts and propositions from the proposed theoretical framework. Third, where no appropriate researches could be found, I have concocted hypothetical studies that, to my knowledge, have never been carried out but are capable of execution. The investigations cited have been drawn from diverse disciplines and reflect a range of theoretical orientations. In addition, I have tried to select researches conducted in or concerned with varied settings (such as homes, hospitals, day care centers, preschools, schools, camps, institutions, offices and factories), contrasting broader social contexts (social classes, ethnic and religious groups, and total societies), and different age levels from early infancy through the life span. Unhappily, these attempts at achieving some representativeness across the spectra of ecology and age met with only partial success. To the extent that they exist, ecologically oriented investigations of development in real-life settings have most often been conducted with infants and preschoolers studied in home or center. Acceptable research designs involving school-age children, adolescents, or adults observed in extrafamilial settings are few.

Having these goals, the volume is admittedly broad in scope. But it is not all-inclusive. No attempt is made to treat the standard subject matter of developmental psychology, that is, to describe the evolution of cognitive, emotional, and social processes over the life course. Nor is particular attention given to a second major preoccupation of contemporary developmental research—the mechanisms of socialization, such as reinforcement and modeling. The omissions do not reflect any lack of interest in these topics. On the contrary, the present work is motivated by my conviction that further advance in the scientific understanding of the basic intrapsychic and interpersonal processes of human development requires their investigation in the actual environments, both immediate and remote, in which human beings live. This task demands the construction of a theoretical schema that will permit the systematic description and analysis of these contexts, their interconnections, and the processes through which these structures and linkages can affect the course of development, both directly and indirectly.

I have thus eschewed the conventional organization of developmental topics in terms either of successive age levels (such as infancy, childhood, and adolescence) or of the classical psychological processes (perception, motivation, learning, and so on). Instead the sections and chapters of this volume reflect the proposed theoretical framework for an ecology of human development. Following a definition of basic concepts, successive chapters deal with elements of the microsystem (chapters 3 through 5), the joint effect of these elements as they function in specific settings (chapters 6 through 8), and the structures and operations of higher order systems at the meso-, exo-, and macro- levels (chapters 9 through 11).

One may well ask how an ecology of human development differs from social psychology on the one hand and sociology or anthropology on the other. In general the answer lies in the focus of the present undertaking on the phenomenon of *development-in-context*. Not only are the above three social science disciplines considerably broader, but none has the phenomenon of development as its primary concern. To describe the ecology of human development as the social psychology, sociology, or anthropology of human development is to overlook the crucial part played in psychological growth by biological factors, such as physical characteristics and in particular the impact of genetic propensities. Indeed the present work does not give such biological influences their due, once again because this cannot be done satisfactorily until an adequate framework for analyzing the environmental side of the equation has been developed, so that the interaction of biological and social forces can be specified.

Finally, lying at the very core of an ecological orientation and distinguishing it most sharply from prevailing approaches to the study of human development is the concern with the progressive accommodation between a growing human organism and its immediate environment, *and* the way in which this relation is mediated by forces emanating from more remote regions in the larger physical and social milieu. The ecology of human development lies at a point of convergence among the disciplines of the biological, psychological, and social sciences as they bear on the evolution of the individual in society.

The primary purpose of detailed discussions of empirical investigations is not to provide an exhaustive analysis of a particular study in terms of either content or method, nor to reach a definitive evaluation of the validity of the findings and their interpretation. To the extent that such assessments are made, they serve as a means to illustrate the practical feasibility, scientific utility, and possible substantive outcomes of an ecological model for the study of human development. Many of the works cited will have conventional virtues or faults that would deserve comment in a more comprehensive treatment but do not bear on the ecological issues under consideration.

Even more disconcerting to the reader may be the fact that many of the studies cited fall short of, or even violate, the principles set forth in this volume, including the very proposition that a given investigation is supposed to illustrate. Such is the present state of the field. I have tried to pick the best examples I could find, but most of them are only partially satisfactory. Rigorous research on human development using ecologically valid measures on both the independent and dependent side of the developmental equation and at the same time paying attention to the influence of larger social contexts is still the exception rather than the rule. At best, one or two important criteria are met, but other features remain at odds with ecological requirements of equal importance. The most typical pattern is one in which the critical conditions are satisfied on one side of the hypothesis but not on the other. For example, an investigation conducted in a real-life setting with systematic description and analysis of relevant physical and social conditions may employ outcome measures, such as an IQ test, a projective technique, or a laboratory procedure, that are of unknown applicability to the environments of scientific interest. Conversely, in another study, the dependent variables may be solidly based in experiences and contexts of everyday life but the independent factors limited to diffuse, dichotomous, and often value-laden labels (middle class and working class, black and white, single-parent family and intact family), with no other contextual evidence provided. The one-sided pattern is so common that to call attention to every instance of its occurrence would be cumbersome. Accordingly, the identification of departures from the requirements of an ecological model is usually limited to violations of principles directly under discussion.

It is important to emphasize in this connection that it is neither necessary nor possible to meet all the criteria for ecological research within a single investigation. Provided the researcher recognizes which qualifications are and are not met, useful scientific information can be gained.

Another shortcoming in the studies cited also reflects the present state of developmental research. I have taken the position that development implies enduring changes that carry over to other places at other times. In the absence of evidence for such carry-over, the observed alteration in behavior may reflect only a short-lived adaptation to the immediate situation. For many of the ideas presented in this volume, it has been impossible to find an example in the research literature that met this important criterion. The great majority of studies in the field of human development do not in fact investigate changes in a person over any considerable time, for they are typically based on brief assessments in a laboratory or testing room that are seldom repeated at a more distant time. One is left to assume that the processes occurring during the original short session will have lasting effect.

Two final disclaimers relate not to the cited researches themselves but to the hypotheses to which they are said to give rise. First, my reasoning may, on occasion, appear somewhat far-fetched. Once again, I merely used the best examples I could find, in the belief that an illustration bearing some relation, however remote, to empirical reality would be preferable to a hypothetical instance.

Second, the justification for this practice is the purpose that hypotheses are intended to serve in the present volume, for they are not offered as definitive propositions. The likelihood that they will be validated in the form in which they are stated is, in my judgment, rather slim. The function of the proposed hypotheses is essentially heuristic—to identify questions, domains, and possibilities believed worthy of exploration.

It is with the aim of contributing to theoretical and empirical discovery that I have written this book. It will have achieved its objective not if the ideas presented prove to be precisely correct, which is improbable, but if their investigation offers new, revealing vistas for the scientific understanding of the forces shaping the development of human beings in the environments in which they live.

2.

Basic Concepts

To assert that human development is a product of interaction between the growing human organism and its environment is to state what is almost a commonplace in behavioral science. It is a proposition that all students of behavior would find familiar, with which none would take issue, and that few would regard as in any way remarkable, let alone revolutionary, in its scientific implications. I am one of those few. I regard the statement as remarkable because of the striking contrast between the universally approved twofold emphasis that it mandates and the conspicuously one-sided implementation the principle has received in the development of scientific theory and empirical work.

To be specific, the principle asserts that behavior evolves as a function of the interplay between person and environment, expressed symbolically in Kurt Lewin's classic equation: $B = f(PE)$ (Lewin, 1935, p. 73). One would therefore expect psychology, defined as the science of behavior, to give substantial if not equal emphasis to both elements on the independent side of the equation, to investigate the person *and* the environment, with special attention to the interaction between the two. What we find in practice, however, is a marked asymmetry, a hypertrophy of theory and research focusing on the properties of the person and only the most rudimentary conception and characterization of the environment in which the person is found.

To appreciate the contrast, one has only to examine the basic texts, books of readings, handbooks, and research journals in psychology in general and developmental psychology in particular. Upon perusing such materials, one will quickly discover concepts and data without end dealing with the qualities of the person. The researcher has available a rich array of personality typologies, de-

16

velopmental stages, and dispositional constructs, each with their matching measurement techniques, that provide highly differentiated profiles of the abilities, temperament, and predominant behavior tendencies of the individual. On the environmental side, however, the prospect is bland by comparison, both in theory and data. The existing concepts are limited to a few crude and undifferentiated categories that do little more than locate people in terms of their social address—the setting from which they come. Thus an examination of studies of environmental influences appearing in a representative sample of texts, books of readings, and journal issues in child psychology and related fields reveals the following modal typologies for describing contexts of behavior and development: family size, ordinal position, single- versus two-parent households, home care versus day care, parents versus peers, and —perhaps the most frequent—variation by social class or ethnic background. Moreover, the data in these studies consist to an overwhelming degree of information not about the settings from which the persons come but about the characteristics of the persons themselves, that is, how people from diverse contexts differ from one another.

As a result, interpretations of environmental effects are often couched in what Lewin called class-theoretical terms; thus observed differences in children from one or another setting (for example, lower class versus middle class, French versus American, day care versus home care) are "explained" simply as attributes of the setting in question. Even when the environment is described, it is in terms of a static structure that makes no allowance for the evolving processes of interaction through which the behavior of participants in the system is instigated, sustained, and developed.

Finally, and perhaps ironically, the data in these studies are typically obtained by removing the research subjects from the particular settings under investigation and placing them in a laboratory or a psychological testing room.[1] The possible impact of these rather special settings on the behavior being elicited, however, is rarely taken into account.

To be sure, there are two spheres of investigation in which some degree of specificity in the analysis of environments is achieved, but the result falls far short of the requirements of an ecological research model. One of these areas, lying primarily in the domain of social psychology, is the study of interpersonal relations and small groups. Given that the people with whom one interacts in a face-to-face situation constitute a part of one's environment, there is a significant body of theory and research dealing with the impact of the environment, in the form of interpersonal influences, on the evolution of behavior. Indeed, to the extent that we have theories about *how* environmental influences affect behavior and development, they are theories about interpersonal processes—reinforcement, modeling, identification, and social learning. From an ecological perspective, such formulations have two shortcomings. First, they tend to overlook the impact of the nonsocial aspects of the environment, including the substantive nature of the activities engaged in by the participants. Second, and more crucial, they delimit the concept of environment to a single immediate setting containing the subject, what in this book is referred to as the *microsystem*. Seldom is attention paid to the person's behavior in more than one setting or to the way in which relations between settings can affect what happens within them. Rarest of all is the recognition that environmental events and conditions outside any immediate setting containing the person can have a profound influence on behavior and development within that setting. Such external influences can, for example, play a critical role in defining the meaning of the immediate situation to the person. Unless this possibility is taken into account in the theoretical model guiding the interpretation of results, the findings can lead to misleading conclusions that both narrow and distort our scientific understanding of the determinants, processes, and potential of human development.

There exists a second body of scholarly work in which external environmental contexts are described in considerable detail and their impact on the course of development graphically traced. Such investigations are carried out primarily in the field of anthropology and to some extent in social work, social psychiatry, clinical psychology, and sociology. But the descriptive material in these studies is heavily anecdotal and the interpretation of causal influences highly subjective and inferential. Here we encounter what I view as an unfortunate and unnecessary schism in contemporary studies of human development. Especially in recent years, research in this sphere has pursued a divided course, each tangential to genuine scientific progress. To corrupt a modern metaphor, we risk being caught between a rock and a *soft* place. The rock is rigor, and the soft place relevance. The emphasis on rigor has led to experiments that are elegantly designed but often limited in scope. This limitation derives from the fact that many of these experiments involve situations that are unfamiliar, artificial, and short-lived, and call for unusual behaviors that are difficult to generalize to other settings. From this perspective, it can be said that much of developmental psychology, as it now exists, is *the science of the strange behavior of children in strange situations with strange adults for the briefest possible periods of time.*[2]

Partially in reaction to such shortcomings, other workers have stressed the need for social relevance in research, but often with indifference to or open rejection of rigor. In its more extreme manifestations, this trend has taken the form of excluding the scientists themselves from the research process. One major foundation has a policy that grants for research will be awarded only to persons who are themselves the victims of social injustice. Less radical expressions of this trend involve reliance on existential approaches in which "experience" takes the place of observation and analysis is forgone in favor of a more personalized and direct "understanding" gained through intimate involvement in the field situation. More common, and more scientifically defensible, is an emphasis on naturalistic observation, but with the stipulation that it be "theoretically neutral" (Barker and Wright, 1954, p. 14), hence unguided by any explicit hypotheses formulated in advance and uncontaminated by highly structured experimental designs imposed prior to data collection.

The most sophisticated argument advocating the superiority of naturalistic over experimental methods in the study of human development emphasizes the practical and ethical impossibility of manipulating and controlling variables of primary significance for psychological growth. For example, in a searching critique of contemporary research in developmental psychology, McCall (1977) starts from a position identical to mine: "It is suggested that, at present, we essentially lack a science of natural developmental processes because few studies are concerned with development as it transpires in naturalistic environments and because we rarely actually collect or analyze truly developmental data. This problem is believed to derive from the veneration of manipulative experimental methods, which have come to dictate rather than serve research questions" (p. 333).

McCall then proceeds to argue that experimental methods, while ideally suited for research in laboratory settings, are ill adapted to the study of "behavior as it typically develops in natural life circumstances" (p. 334), since, for practical and ethical reasons, it is impossible to manipulate and control all the revelant factors. In McCall's words,

There is nothing inherently wrong with manipulative experimental studies in developmental psychology, but this methodology . . . is often impossible to execute . . . For example, exposure to visual pattern is required for the development of a variety of visual functions, but every child receives adequate patterned light. Certain sensorimotor activities may be propaedeutic to the acquisition of agent-action-object language constructions, but almost all children obtain adequate amounts of these experiences . . . To determine the necessary causes of development, one must deprive the organism of the hypothetical circumstance. However, when children are the focus of study, ethical considerations obviate experimental deprivation in most cases.

We must simply accept the fact from logical and practical standpoints that we will probably never prove the sufficient or necessary cause for the naturalistic development of a host of major behaviors, some of which represent the essence of our discipline. (Pp. 335–336)

McCall's persuasive argument assumes that the only function of the experiment in science is to establish necessary and sufficient conditions. As I argue later, to make this assumption is seriously to underestimate the scientific power of the experimental method: the experimental method is not only invaluable for the verification of hypotheses; it is equally and perhaps even more applicable to their discovery. In short, for science in general and especially for rigorous research on development-in-context, the experiment is a powerful and essential heuristic tool.

For these reasons, the orientation proposed here rejects both the implied dichotomy between rigor and relevance and the assumed incompatibility between the requirements of research in natural situations and the applicability of structured experiments at an early stage in the scientific process. It rejects as spurious the argument that, because naturalistic observation preceded experimentation in both the physical and the biological sciences, this progression is necessarily the strategy of choice in the study of human behavior and development. Such an interpretation mistakes a historical sequence for a causal one and represents yet another instance of the logical pitfalls inherent in the ever seductive *post hoc,* ergo *propter hoc* inference. In my view, twentieth-century science possesses research strategies that, had they been available to the nineteenth-century naturalists, would have enabled them to leapfrog years of painstaking, exhaustive description in arriving at a formulation of biological principles and laws. This is not to imply that taxonomy is not an essential scientific task but only to assert that a phase of purely descriptive observation, recording, and classification may not be a necessary condition for making progress in the understanding of process and that the early application of experimental paradigms may in fact lead to more appropriate taxonomies for achieving the requisite work of the systematic ordering of natural phenomena.

Yet another restriction is unnecessarily imposed on the strategy of naturalistic observation, particularly as applied to the human case by its principal advocates—the ethnologists (Jones, 1972; McGrew, 1972) and the psychological ecologists of the Kansas school (Barker and Schoggen, 1973; Barker and Wright, 1954). Both groups have adapted to the study of human behavior a model originally developed for the observation of subhuman species. Implicit in this model is a concept of the environment that may be quite adequate for the study of behavior in animals but that is hardly sufficient for the human case: it is limited to the immediate, concrete setting containing the living creature and focuses on the observation of the behavior of one or, at most, two beings at a time in only one setting. As I shall argue below, the understanding of *human* development demands more than the direct observation of behavior on the part of one or two persons in the same place; it requires examination of multiperson systems of interaction not limited to a single setting and must take into account aspects of the environment beyond the immediate situation containing the subject. In the absence of such a broadened perspective, much of contemporary research can be characterized as the study of *development-out-of-context*.

The present work offers a foundation for building context into the research model at the levels of both theory and empirical work. I propose first an expansion and then a convergence of the naturalistic and the experimental approaches—more precisely, an expansion and convergence of the theoretical conceptions of the environment that underlie each of them. I refer to this evolving scientific perspective as the *ecology of human development*.

I begin with some definitions of substantive focus.

DEFINITION 1

The ecology of human development involves the scientific study of the progressive, mutual accommodation between an active, growing human being and the changing properties of the immediate settings in which the developing person lives, as this process is affected by relations between these settings, and by the larger contexts in which the settings are embedded.

Three features of this definition are especially worthy of note. First, the developing person is viewed not merely as a tabula rasa on which the environment makes its impact, but as a growing, dynamic entity that progressively moves into and restructures the milieu in which it resides. Second, since the environment also exerts its influence, requiring a process of mutual accommodation, the interaction between person and environment is viewed as two-directional, that is, characterized by *reciprocity*. Third, the environment defined as relevant to developmental processes is not limited to a single, immediate setting but is extended to incorporate interconnections between such settings, as well as to external influences emanating from the larger surroundings. This extended conception of the environment is considerably broader and more differentiated than that found in psychology in general and in developmental psychology in particular. The *ecological environment* is conceived topologically as a nested arrangement of concentric structures, each contained within the next. These structures are referred to as the *micro-, meso-, exo-,* and *macrosystems,* defined as follows.

DEFINITION 2
A microsystem is a pattern of activities, roles, and interpersonal relations experienced by the developing person in a given setting with particular physical and material characteristics.

A *setting* is a place where people can readily engage in face-to-face interaction—home, day care center, playground, and so on. The factors of *activity, role,* and *interpersonal relation* constitute the *elements,* or building blocks, of the microsystem.

A critical term in the definition of the microsystem is *experienced.* The term is used to indicate that the scientifically relevant features of any environment include not only its objective properties but also the way in which these properties are perceived by the persons in that environment. This emphasis on a phenomenological view springs neither from any antipathy to behavioristic concepts nor from a predilection for existential philosophical foundations. It is dictated simply by a hard fact. Very few of the external influences significantly affecting human behavior and development can be described solely in terms of objective physical conditions and events; the aspects of the environment that are most powerful in shaping the course of psychological growth are overwhelmingly those that have meaning to the person in a given situation.

There is, of course, nothing original in this formulation. It draws heavily on the work of theorists from a variety of disciplines. From philosophy and psychology, it builds on the phenomenological concepts of Husserl (1950), Köhler (1938), and Katz (1930). In sociology, an analogous formulation has its roots in the role theory of George Herbert Mead (1934), and is epitomized in the Thomases' concept of the "definition of the situation" (Thomas and Thomas, 1928). In psychiatry, the view was brilliantly applied to the study of interpersonal relations and psychopathology by Sullivan (1947). In education, the orientation is found in Dewey's emphasis on designing curricula that reflect the everyday experience of the child (1913, 1916, 1931). In anthropology, the approach has been extended to the analysis of larger social systems, most notably by Linton (1936) and Benedict (1934). Its significance for the general study of human behavior is summed up in what is perhaps the only proposition in social science that approaches the status of an immutable law—the Thomases' inexorable dictum "If men define situations as real, they are real in their consequences" (Thomas and Thomas, 1928, p. 572).

In the main, however, the phenomenological conception of the environment that lies at the foundation of the theory derives its structure and rationale from the ideas of Kurt Lewin, especially his construct of the "life space" or "psychological field" (1931, 1935, 1951). Lewin takes the position that the environment of greatest relevance for the scientific understanding of behavior and development is reality not as it exists in the so-called objective world but as it appears in the mind of the person; in other words, he focuses on the way in which the environment is perceived by the human beings who interact within and with it. An especially significant aspect of this perceived environment is the world of imagination, fantasy, and unreality. Yet despite such seeming richness, Lewin's theoretical map of the psychological field is curiously lacking in content. To use his own term, his is a "topological psychology," a systematic description of a space without substance, replete with empty regions and nested structures, separated by boundaries, joined by interconnections and pathways, and beset by barriers and detours on the way to unspecified goals. The most unorthodox aspect of Lewin's schema is his treatment of motivational forces as emanating not from within the person but from the environment itself. Objects, activities, and especially other people send out lines of force, valances, and vectors that attract and repel, thereby steering behavior and development.

What could all this mean in concrete terms? What sense, let alone application, could one make of a theory in which the perceived is viewed as more important than the actual, the unreal more valid than the real; where the motivation that steers behavior inheres in external objects, activities, and other people; and where the content of all these complicated structures remains unspecified? More pointedly, how could anyone apply such airy abstractions to settings in everyday life, or for that matter, why should anyone wish to do so?

A basis for a plausible answer to these questions is suggested by consideration of the very first paper Lewin wrote, "Kriegsland-schaft" ("War Landscape"), published at the end of the First World War after he had spent several years in the army, most of it in the front lines where he had been wounded in combat. The article, which appeared in the *Zeitschrift für Angewandte Psychologie* (1917), represents a marvelous prefiguring of all his basic theoretical concepts. In this extraordinary paper, Lewin describes how the perceived reality of the landscape changes as one moves nearer to the front. What first appears as a lovely bucolic scene of farmhouses, fields, and wooded areas is gradually transformed. The forested hill-top becomes an observation post, its sheltered side the location for a gun emplacement. An unexposed hollow is seen as a probable battalion aid station. Aspects of the natural landscape that were a delight only a few kilometers back are now perceived as ominous: the frightening defile, the camouflage of trees, the hill that hides the unseen enemy, the invisible objective to be taken, the place and moment of security after the fray—features of the environment that threaten, beckon, reassure, and steer one's course across a terrain objectively undistinguishable from scenes just a short distance behind the front.

Here are the basic premises of what later became Lewin's explicit, systematic theory: the primacy of the phenomenological over the real environment in steering behavior; the impossibility of understanding that behavior solely from the objective properties of an environment without reference to its meaning for the people in the setting; the palpable motivational character of environmental objects and events; and, especially, the importance of the unreal, the imagined—the enemy not seen, the promise of a warm meal, and the prospect of surviving to sleep, or to lie awake another night. What could be more down to earth than this?

Herein also lies the explanation for Lewin's unwillingness to specify in advance the content of the psychological field: it is a terrain that has yet to be explored. Such exploration, therefore, constitutes a major task of psychological science. One needs to discover empirically how situations are perceived by the people who participate in them. Again, without specifying content, Lewin distinguishes two aspects of every situation that are likely to capture the person's attention. The first is Tätigkeit, perhaps best translated as "ongoing activity"; it refers to the tasks or operations in which a person sees himself or others as engaging. The second salient feature involves the perceived interconnections between the people in the setting, in terms not so much of interpersonal feelings as of the relations of the various parties with each other as members of a group engaged in common, complementary, or relatively independent undertakings.

In addition to these two aspects of the situation highlighted by Lewin, the concept of microsystem involves a third feature emphasized in the sociological theories of Mead and the Thomases, namely, the notion of role. For the present, we can make use of the standard definition of role in the social sciences: a set of behaviors and expectations associated with a position in society, such as that of mother, baby, teacher, friend, and so on.

The phenomenological perspective is also relevant at the next and succeeding levels of ecological structure.

DEFINITION 3

A mesosystem comprises the interrelations among two or more settings in which the developing person actively participates (such as, for a child, the relations among home, school, and neighborhood peer group; for an adult, among family, work, and social life).

A mesosystem is thus a system of microsystems. It is formed or extended whenever the developing person moves into a new setting. Besides this primary link, interconnections may take a number of additional forms: other persons who participate actively in both settings, intermediate links in a social network, formal and informal communications among settings, and, again clearly in the phenomenological domain, the extent and nature of knowledge and attitudes existing in one setting about the other.

DEFINITION 4

An exosystem refers to one or more settings that do not involve the developing person as an active participant, but in which events occur that affect, or are affected by, what happens in the setting containing the developing person.

Examples of an exosystem in the case of a young child might include the parent's place of work, a school class attended by an older sibling, the parents' network of friends, the activities of the local school board, and so on.

DEFINITION 5
The *macrosystem* refers to consistencies, in the form and content of lower-order systems (micro-, meso-, and exo-) that exist, or could exist, at the level of the subculture or the culture as a whole, along with any belief systems or ideology underlying such consistencies.

For example, within a given society—say France—one crèche, school classroom, park playground, café, or post office looks and functions much like another, but they all differ from their counterparts in the United States. It is as if in each country the various settings had been constructed from the same set of blueprints. An analogous difference in form appears at levels beyond the microsystem. Thus the relations between home and school are rather different in France than in our own country. But there are also consistent patterns of differentiation within each of these societies. In both worlds, homes, day care centers, neighborhoods, work settings, and the relations between them are not the same for well-to-do families as for the poor. Such intrasocietal contrasts also represent macrosystem phenomena. The systems blueprints differ for various socioeconomic, ethnic, religious, and other subcultural groups, reflecting contrasting belief systems and lifestyles, which in turn help to perpetuate the ecological environments specific to each group.

I deliberately mention in the definition of macrosystem patterns that "could exist" so as to expand the concept of macrosystem beyond limitation to the status quo to encompass possible blueprints for the future as reflected in the vision of a society's political leaders, social planners, philosophers, and social scientists engaging in critical analysis and experimental alteration of prevailing social systems.

Having been introduced to the structure of the ecological environment, we are now in a position to identify a general phenomenon of movement through ecological space—one that is both a product and a producer of developmental change.

DEFINITION 6
An ecological transition occurs whenever a person's position in the ecological environment is altered as the result of a change in role, setting, or both.

Instances of ecological transition as defined here occur throughout the life span. To name but a few: a mother is presented with her newborn infant for the first time; mother and baby return home from the hospital; there is a succession of baby sitters; the child enters day care; a younger sibling arrives; Johnny or Mary goes to school, is promoted, graduates, or perhaps drops out. Then there is finding a job, changing jobs, losing jobs; marrying, deciding to have a child; having relatives or friends move in (and out again); buying one's first family car, television set, or home; vacationing, traveling; moving; getting divorced, remarrying; changing careers; emigrating; or, to turn to even more universal themes: becoming sick, going to the hospital, getting well again; returning to work, retiring; and the final transition to which there are no exceptions—dying.

I shall argue that every ecological transition is both a consequence and an instigator of developmental processes. As the examples indicate, the transitions are a joint function of biological changes and altered environmental circumstances; thus they represent examples par excellence of the process of mutual accommodation between the organism and its surroundings that is the primary focus of what I have called the ecology of human development. Furthermore, the alterations in the milieu can occur at any of the four levels of the ecological environment. The appearance of a younger sibling is a microsystem phenomenon, entry into school changes exo- into mesosystem, and emigrating to another country (or perhaps just visiting the home of a friend from a different socioeconomic or cultural background) involves crossing macrosystem borders. Finally, from the viewpoint of research, every ecological transition constitutes, in effect, a ready-made experiment of nature with a built-in, before-after design in which each subject can serve as his own control. In sum, an ecological transition sets the stage both for the occurrence and the systematic study of developmental phenomena.

We are brought back to the fundamental question of how development is to be conceived in the framework of an ecological theory. The formulation presented here starts from the proposition that development never takes place in a vacuum; it is always embedded and expressed through behavior in a particular environmental context.

DEFINITION 7
Human development is the process through which the growing person acquires a more extended differentiated, and valid conception of the ecological environment, and becomes motivated and able to engage in activities that reveal the properties of, sustain, or restructure that environment at levels of similar or greater complexity in form and content.

Three features of this definition are particularly worthy of note. First, development involves a change in the characteristics of the person that is neither ephemeral nor situation-bound; it implies a reorganization that has some continuity over both time and space. Second, developmental change takes place concurrently in two domains, those of perception and action. Third, from a theoretical viewpoint, each of these domains has a structure that is isomorphic with the four levels of the ecological environment. Thus in the perceptual sphere the question becomes to what extent the developing person's view of the world extends beyond the immediate situation to include a picture of other settings in which he has actively participated, the relations among these settings, the nature and influence of external contexts with which he has had no face-to-face contact, and, finally, the consistent patterns of social organization, belief systems, and lifestyle specific to his own and other cultures and subcultures. Analogously, at the level of action, at issue is the person's capacity to employ strategies that are effective, first, in providing accurate feedback about the nature of the systems existing at successively more remote levels, second, enabling these systems to continue to function and, third, reorganizing existing systems or creating new ones of comparable or higher order that are more in accord with his desires. Later I shall endeavor to show how this two-sided ecological conception of development can be fruitfully applied both to obtain a richer scientific yield from existing research findings and to design new investigations that will further illuminate the nature, course, and conditions of human development.

An ecological conception of *development-in-context* also has implications for research method and design. To begin, it accords key importance to and provides the theoretical basis for a systematic definition of a construct often alluded to in recent discussions of developmental research—*ecological validity*. Although the term as yet has no accepted definition, one can infer from these discussions a common underlying conception: an investigation is regarded as ecologically valid if it is carried out in a natural setting and involves objects and activities from everyday life. Although originally attracted to this notion, I have upon reflection come to view it not only as too simplistic but as scientifically unsound on several counts. First, while I agree wholeheartedly with the desirability of extending research activities beyond the laboratory, I question the seemingly automatic granting of scientific legitimacy to a research effort merely because it is conducted in a real-life setting. Even more arbitrary is the converse implication that any investigation carried out in a nonnatural setting is necessarily ecologically invalid and thereby scientifically suspect on purely a priori grounds. Surely this is to prejudge the issue. Moreover, the term *ecological validity,* as it is currently used, has no logical relation to the classical definition of validity—namely, the extent to which a research procedure measures what it is supposed to measure. Indeed, there is a basic conflict between the theoretical assumptions underlying the two conceptions. In the classical definition, validity is ultimately determined by the nature of the problem under investigation. By contrast, ecological validity as heretofore defined appears to be determined once and for all by the setting in which the study is conducted, without regard to the question under investigation. In any research endeavor this last consideration must be the most decisive in assessing validity of whatever kind.

At the same time, there is implicit in current concerns with ecological validity another principle that can no longer be disregarded in the light of available evidence. This is the proposition that the properties of the environmental contexts in which an investigation is conducted or from which the experimental subjects come can influence the processes that take place within the research setting and thereby affect the interpretation and generalizability of the findings.

I have therefore sought to formulate a definition of ecological validity that takes both these principles into account. Once the task was undertaken, it was not difficult to achieve. All that was required was a logical extension of the traditional definition of validity. This definition is limited in focus, applying only to the measurement procedures employed in research operations. The definition of ecological validity proposed here expands the scope of the original concept to include the environmental context in which the research is conducted.

DEFINITION 8
Ecological validity refers to the extent to which the environment experienced by the subjects in a scientific investigation has the properties it is supposed or assumed to have by the investigator.

Again, the use of the term *experienced* in the definition highlights the importance of the phenomenological field in ecological research. The ecological validity of any scientific effort is called into question whenever there is a discrepancy between the subject's perception of the research situation and the environmental conditions intended or assumed by the investigator. This means that it becomes not only desirable but essential to take into account in every scientific inquiry about human behavior and development how the research situation was perceived and interpreted by the subjects of the study. The importance of this injunction will become apparent when, later in this volume, we examine specific investigations from the perspective of ecological validity and find ourselves arriving at plausible alternative interpretations that cannot be resolved without our having *at least some* knowledge of the subject's definition of the situation.

In one of the few systematic analyses of the concept of ecological validity, Michael Cole and his colleagues (1978) point out that the task of determining how the subject perceives the situation is an extremely difficult one that the psychological researcher does not yet know how to accomplish. They go on to argue that Lewin's emphasis on this requirement as central to ecological validity (1943) is difficult to reconcile with the scientific demands of an alternative formulation of the concept proposed by Lewin's contemporary, Egon Brunswik (1943, 1956, 1957). Brunswik used the term in a far narrower sense to apply to a more traditional problem in the psychology of perception—the relation between a proximal cue and the distal object in the environment to which it was related. The ecological element in this conception derived from Brunswik's insistence on "representative design." In his view, to establish the existence of a given psychological process it was necessary to demonstrate its occurrence across a sample not only of subjects but also of situations. The purpose of such environmental sampling was to show that the phenomenon "possesses generality with regard to normal life conditions" (1943, p. 265).

While applauding Brunswik's emphasis on the importance of conditions of everyday life as proper referents for basic research, I shall later (chapter 6) take issue with the fundamental assumption underlying Brunswik's argument, and much of contemporary psychological science as well, that the only processes meriting scientific status in the study of human behavior are those that are invariant across contexts. For the moment, however, our concern is with the contention of Cole and company that, in practice if not in theory, the ecological requirements of Lewin and of Brunswik are incompatible with each other. They claim that to insist that research be carried out in a variety of situations and, at the same time, demand that each situation be examined in terms of its psychological meaning to the participants imposes "an enormous burden" on the investigator, one that "is perhaps more than psychology can, or psychologists would care to take on" (Cole, Hodd, and McDermott, 1978, p. 36).

The charge is a serious one and deserves a serious answer. A first response does not resolve the dilemma, but only reaffirms that it is unavoidable. To disregard the meaning of the situation to the research subject is to risk invalid conclusions both for research and, particularly in the study of human development, for public policy. To close one's eyes to this possibility is, therefore, to be scientifically and socially irresponsible. But how is one to deal with the dilemma posed by Cole and his colleagues? Ironically, one approach to resolution is found in the work of Cole himself. In two important volumes (Cole and Scribner, 1974; Cole et al., 1971) he and his associates develop the position that the significance of much of the behavior taking place in a given social setting *can* be understood, *provided* the observer has participated in the given setting in roles similar to those taken by the participants and is a member of or has had extensive experience in the subculture in which the setting occurs and from which the actors come. This proviso still leaves much room for misconception, but it considerably reduces the likelihood of gross errors of misinterpretation. The situation is analogous to that faced by a person doing simultaneous translation at an international meeting. To accomplish the task, it is helpful—but not absolutely essential—to be a native speaker; it is a sine qua non, however, to be experienced in the ways of international conferences, have good knowledge of the subject matter, and possess full command of both languages.

The nature and necessity of these requirements is obvious enough in the case of simultaneous translation. Moreover, they are scrupulously adhered to, primarily because the participants in the proceedings have access to the record, and possess the power to press for its correction. The situation is somewhat different for the researcher of human behavior. In that case, the requirements are more one-sided: the emphasis is on mastering the knowledge, technology, and language of science rather than of the settings or persons under study. Indeed, the latter are seldom informed about the content of the scientific record and have no power to alter it. In the absence of persons able to recognize unwarranted interpretations based on misperceptions of fact, the unwitting investigator can, in all good faith, arrive at false conclusions. Once such persons are involved in the scientific enterprise, the risk of errors is appreciably reduced.

The involvement of people from the subject's world in the research process implies a significant reorientation in the traditional relation between the researcher and the researched in the behavioral sciences. As reflected in the classical experimenter-subject paradigm of the laboratory, the former is typically thought of by both parties as possessing greater knowledge and control, whereas the latter is asked, and expected, to accept the situation as structured and to cooperate in acting as requested. An ecological orientation emphasizing the subject's definition of the situation accords far more importance to the knowledge and initiative of the persons under study. Experimental instructions and manipulations are by no means ruled out but are directed toward clarifying or determining the objective features of the environment (for example, selecting the setting, allocating roles, assigning tasks) rather than specifying the particular ways in which the subject is to behave. For by allowing activities to emerge spontaneously within the given environmental context, the investigator can obtain evidence bearing on the psychological meaning of the context to the participants.

There are of course other strategies for probing the content of the psychological field. They include interviewing participants after the fact to discover whether their retrospective view of the situation is consistent with the intention of the investigator, as well as introducing the same activities into different settings (for example the home and the laboratory) to identify any systematic effects of context.

But even if all these measures are taken, even if observers are fully familiar with the setting and the subculture, the research situation structured so as to give relatively free rein to activities initiated by the participants, the latter given opporunity to examine and comment on the scientific results and their interpretation, and investigations conducted in different contexts to highlight the distinctive features of particular settings—even if all this is achieved, serious problems still remain in ascertaining how the research situation was perceived by the persons under study. Particularly in developmental research, there exists the intriguing and often insoluble problem of understanding the phenomenological world of the infant and the young child before they can provide glimpses of their psychological experience through language. Even with adults, there is the inevitable phenomenon of idiosyncratic perception based on past experience and internal states hidden from the observer.

It was undoubtedly considerations such as these that led Cole and his associates to come to a determined but carefully qualified stance regarding the importance and feasibility of establishing phenomenologically based ecological validity in their own sphere of special interest—cognitive development. In the final paragraph of their analysis, they offer this sobering conclusion:

We need to know as much as possible about the subject's responses to the task-as-posed, because this is crucial information for both Brunswik's and Bronfenbrenner's notions of ecological validity. There are no currently agreed-upon methods for accomplishing these goals. While several investigators, including ourselves, are engaged in the required methods, claims for the ecological validity of cognitive tasks should be treated as programmatic hopes for the future. We have made little progress on this issue since Brunswik's and Lewin's discussion a generation ago. (1978, p. 37)

Along with the work of Cole and his associates, the present volume represents an attempt to move the field a step beyond Brunswik's and Lewin's pioneering ideas by offering a conceptual framework for analyzing the psychological life space in terms of the three microsystem elements of activity, role, and relation. The effort may not take us very far, but any added information about the nature of the perceived environment is a scientific gain in the study of development-in-context. Herein lies the basis for a somewhat more optimistic interpretation of the operational dilemma correctly posed by Cole and his associates, for it is neither necessary nor even possible to obtain a *complete* picture of the research situation as perceived by the participants. Like frictionless motion, ecological validity is a goal to be pursued, approached, but never achieved. The more closely it is approximated, however, the clearer will be the scientific understanding of the complex interplay between the developing human organism and the functionally relevant aspects of its physical and social environment.

The scope of this interplay serves as a reminder that correspondence between the subject's and the investigator's view of the research situation, or what might appropriately be called *phenomenological validity*, is only one aspect of ecological validity. Errors of interpretation may also arise because of the investigator's failure to take into account the full range of environmental forces that are operative in a given situation, including those emanating from contexts beyond the immediate setting containing the research subjects—influences at the level of meso-, exo-, and macrosystems.

The notion of ecological validity that I have set forth can be regarded as implicit in the classical definition of scientific validity, since the failure to recognize discrepancies between the subject's and researcher's definition of the situation or the operation of influences from outside the research setting ultimately calls into question whether a given scientific procedure is measuring what it is supposed to measure. The argument follows logically enough. The question is whether its exacting implications will in fact be recognized and heeded in the absence of an explicit requirement to take into account environmental influences, real or perceived, that can affect the validity of research operations. It is this consideration that dictates the necessity of specifying a criterion of ecological validity.

Finally, this definition does not designate any particular kind of research locale as valid or invalid on a priori grounds. Thus depending on the problem, the laboratory may be an altogether appropriate setting for an investigation, and certain real-life environments may be highly inappropriate. Suppose one is interested in studying the interaction between mother and child when the child is placed in a strange and unfamiliar situation. It is clear that the laboratory approximates this condition far better than the home. Conversely, if the focus of inquiry is the modal pattern of parent-child activity prevailing in the family, observations confined to the laboratory can be misleading. As I indicate in chapter 6, findings from a number of studies demonstrate that patterns of parent-child interaction in the home can be substantially and systematically different from those observed in the laboratory. Once again, however, the fact that research results obtained in the laboratory differ from those observed in the home cannot be interpreted as evidence for the superiority of one setting over the other, except in relation to a specific research question. At the very least, such differences serve to illuminate the special properties of the laboratory as an ecological context. More important, they illustrate the as yet unexploited power of the laboratory as an ecological contrast that can highlight the distinctive features of other types of settings as they affect behavior and development. From this point of view, an ecological orientation increases rather than reduces opportunities for laboratory research by pointing to new knowledge that can be achieved through close and continuing interaction between laboratory and field research.

At a more general level, the comparison of results obtained in laboratory and real-life settings provides an illustration of the basic strategy through which ecological validity can be demonstrated or found wanting. As was true for the definition of that concept, the method represents an extension of the procedures employed for investigating validity in its classical form. The process is essentially one of establishing construct validity (Cronbach and Meehl, 1955), in this instance by testing the ecological theory underlying the research operations—the assumptions being made about the nature and generality of the environment in which the research is being conducted. When a laboratory study is regarded as representative of behavior elsewhere, evidence must be provided of an empirical relation to similar activities in the other setting; in other words, validation must take place against an external ecological criterion, with the possibility of systematic divergence explicitly taken into account. It should be recognized, moreover, that such divergence may take the form of differences not merely in average response but in the total pattern of relationships, and in the underlying processes that they are presumed to reflect.

In research on the ecology of human development, the ability to generalize across settings is important for yet another reason. Even after ecological validity has been established, still another criterion must often be met: whenever the hypothesis under investigation implies, as it frequently does, that development has actually occurred, it is necessary to provide evidence of such an outcome before the hypothesis can be regarded as receiving empirical support. As I emphasized earlier, development implies a change that is not merely momentary or situation-specific. It is therefore not sufficient to show only that a certain variation in the environment has produced an alteration in behavior; it is also necessary to demonstrate that this change exhibits some invariance across time, place, or both. We refer to such a demonstration as the establishment of *developmental validity*, defined as follows.

DEFINITION 9
To demonstrate that human development has occurred, it is necessary to establish that a change produced in the person's conceptions and/or activities carries over to other settings and other times. Such demonstration is referred to as developmental validity.

Even the most cursory examination of published research in human development reveals that this principle is honored more in the breach than in the observance. Particularly in laboratory studies, investigations purporting to demonstrate a developmental effect frequently offer in evidence only data that are confined to a single setting and a relatively brief period of time.

As should be true of any scientific endeavor, decisions regarding research design are dictated by theoretical considerations. Given a complex conception of person-environment interaction in the context of interdependent, nested systems, the question arises how these interdependencies can be investigated empirically. I shall argue that a strategy especially well suited for this purpose, from the earliest stages of research forward, is an *ecological experiment*, defined as follows.

DEFINITION 10
An ecological experiment is an effort to investigate the progressive accommodation between the growing human organism and its environment through a systematic contrast between two or more environmental systems or their structural components, with a careful attempt to control other sources of influence either by random assignment (planned experiment) or by matching (natural experiment).

I deliberately eschew the term *quasi-experiment*, typically employed in the research literature, because it suggests a lower level of methodological rigor, an implication I regard as unwarranted on strictly scientific grounds. There are instances in which a design exploiting an experiment of nature provides a more critical contrast, insures greater objectivity, and permits more precise and theoretically significant inferences—in short, is more elegant and constitutes "harder" science—than the best possible contrived experiment addressed to the same research question.

In other respects the definition has a familiar ring. In keeping with the commitment to rigor affirmed at the outset, the main body of the definition is a restatement of the basic logic of the experimental method. What is novel, and perhaps debatable, in this formulation is not the procedure advocated but the timing and the target of its application. I am proposing that experiments be employed in the very first phases of scientific inquiry not for the usual objective of testing hypotheses (although this device is used as a means to an end) but for *heuristic purposes*—namely, to analyze systematically the nature of the existing accommodation between the person and the milieu.

The need for early experimentation derives from the nature of the problem under investigation. The "accommodation" or "fit" between person and environment is not an easy phenomenon to recognize. Here, looking is usually not enough. As Goethe wrote with his poet's prescience, "Was ist das Schwerste von allem? Was dir das Leichste dünket, mit den Augen zu sehen, was vor den Augen dir liegt." (What is the most difficult of all? That which seems to you the easiest, to see with one's eyes what is lying before them.) (*Xenien aus dem Nachlass* #45.)

If looking is not enough, what is one to do? How can the observer quicken his sensitivity to the critical features of the observed? The answer to the question was given me forty years ago, long before I was ready to appreciate it, by my first mentor in graduate school, Walter Fenno Dearborn. In his quiet, crisp New England accent, he once remarked, "Bronfenbrenner, if you want to understand something, try to change it." And whether one studies change by deliberately altering conditions in a contrived experiment or by systematically exploiting an "experiment of nature," the scientific purpose and effect are the same; to maximize one's sensitivity to phenomena through the juxtaposition of the similar but different constitutes the core of the experimental method and creates its magnifying power.

The case presented here for early and continuing application of experimental paradigms should not be misinterpreted as an argument against the use of other methods, such as ethnographic description, naturalistic observation, case studies, field surveys, and so on. These strategies can provide invaluable scientific information and insights. The point being made is a positive one—that the experiment plays a critical role in ecological investigation not only for the purpose of testing hypotheses but, at prior stages, for detecting and analyzing systems properties within the immediate setting and beyond. The special suitability of the experiment for this purpose is highlighted by an adaptation of Dearborn's dictum to the ecological realm: If you wish to understand the relation between the developing person and some aspect of his environment, try to budge the one, and see what happens to the other. Implicit in this injunction is the recognition that the relation between person and environment has the properties of a system with a momentum of its own; the only way to discover the nature of this inertia is to try to disturb the existing equilibrium.

It is from this perspective that the primary purpose of the ecological experiment becomes not hypothesis testing but *discovery*—the identification of those systems properties and processes that affect and are affected by the behavior and development of the human being. Moreover, if the objective is the identification of systems properties, then it is essential that such systems properties not be excluded from research design before the fact by restricting observation to only one setting, one variable, or one subject at a time. Human environments and, even more so, the capacities of human beings to adapt and restructure these environments, are so complex in their basic organization that they are not likely to be captured through simplistic, unidimensional research models that make no provision for assessing ecological structure and variation. Unlike the classical laboratory experiment in which one focuses on a single variable at a time and attempts to "control out" all others, in ecological research the investigator seeks to "control in" as many theoretically relevant ecological contrasts as possible within the constraints of practical feasibility and rigorous experimental design. Only in this way can one assess the generality of a phenomenon beyond a specific ecological situation and, equally significant from a developmental perspective, identify the processes of mutual accommodation between a growing organism and its changing surroundings. For instance, in studying socialization strategies one might do well to stratify the sample not only, as is commonly done, by social class, but also by family structure and/or child-care setting (home versus center care). Such stratification in terms of two or more ecological dimensions provides a systematically differentiated and thereby potentially sensitive grid that makes possible the detection and description of patterns of organism-environment interaction across a range of ecological contexts. Moreover, given the extraordinary capacity of the species homo sapiens to adapt to its milieu, these patterns are more likely to be complex than simple. To corrupt, somewhat, the classical terminology of experimental design, *in ecological research, the principal main effects are likely to be interactions.*

A line of argument that urges the execution of research in more than one setting, as well as multiple classification by ecological categories both within and across settings, invites the counterargument that it is impractical in terms of the magnitude of the undertaking and the number of subjects required. Thus a critic might contend that, under such circumstances, research on the ecology of human development could be conducted only in large-scale projects far beyond the human and material resources ordinarily available to most established scientists, let alone younger investigators and graduate students. While some large-scale studies are indeed desirable, they have no necessary relation to the research model advocated here. It is not the size but the structure of the design that is critical. For instance, research on ecological transitions—such as the effect on the child of the arrival of a sibling, changes in behavior at home as a function of the child's entry into and progress in school, the adaptation of an adolescent to a new father, or the impact on the family of parental unemployment—by no means requires a large number of subjects and could readily be carried out by graduate students or even undergraduate majors, especially if they worked in collaboration. Furthermore, stratification does not necessarily demand the addition of more subjects but only a systematic recognition of the different ecological contexts from which research subjects come and a deliberate selection to insure that at least the most critical and unavoidable contrasts are represented systematically rather than left to chance. Allowing the latter to occur unheeded not only inflates experimental error but also may deprive the investigator of information bearing on the interaction of different ecological conditions in shaping the course of development. The loss in degrees of freedom associated with stratification is, I suggest, more than compensated for by the gain in knowledge about combinatorial contextual effects. The occurrence of such interactions and their significance for science and social policy are illustrated by the results of specific studies reviewed in the chapters that follow. A number of these are small-scale investigations conducted by a single researcher.

I have emphasized the scientific importance of conducting ecological experiments on environmental influences beyond the immediate setting containing the developing person. Especially powerful in this regard are investigations that address properties of the macrosystem. There are two major strategies for investigating the consistent patterns of development-in-context that characterize particular cultures and subcultures. The first is the comparison of existing groups, as exemplified by the large number of studies of socioeconomic and ethnic differences in child-rearing practices and behavior. But since most of these researches focus on the characteristics of individuals almost to the exclusion of the properties of the social contexts in which the individuals are found, they can shed little light on the process of accommodation between person and environment which constitutes the core of an ecology of human development. There are some notable exceptions to this restricted perspective, but even these more broadly conceived investigations share with all strictly naturalistic studies the disadvantage of being limited to variations in macrosystems that presently exist or have occurred in the past. Future possibilities remain uncharted, except by hazardous extrapolation.

This restriction of interest to the status quo represents a distinctive characteristic of much American research on human development. This foreshortened theoretical perspective was first brought to my attention by Professor A. N. Leontiev of the University of Moscow. At the time, more than a decade ago, I was an exchange scientist at the Institute of Psychology there. We had been discussing differences in the assumptions underlying research on human development in the Soviet Union and in the United States. In summing up his views, Professor Leontiev offered the following judgment: "It seems to me that American researchers are constantly seeking to explain how the child came to be what he is; we in the U.S.S.R. are striving to discover not how the child came to be what he is, but how he can become what he not yet is."

Leontiev's statement is of course reminiscent of Dearborn's injunction ("If you want to understand something, try to change it."), but it goes much further; indeed, in Leontiev's view, it is revolutionary in its implications. Soviet psychologists often speak of what they call the "transforming experiment." By this they mean an experiment that radically restructures the environment, producing a new configuration that activates previously unrealized behavioral potentials of the subject. Russian developmental psychologists have indeed been ingenious in devising clever experiments that evoke new patterns of response, primarily in the sphere of psychomotor and perceptual development (Cole and Maltzman, 1969). But once Soviet research moves out of the laboratory, the control group disappears, systematic data yield to anecdotal accounts, and the transforming experiment all to often degenerates into dutiful demonstration of ideologically prescribed processes and outcomes.

For rather different reasons, transforming experiments in the real world are equally rare in American research on human development. As Leontiev implied, most of our scientific ventures into social reality perpetuate the status quo; to the extent that we include ecological contexts in our research, we select and treat them as sociological givens rather than as evolving social systems susceptible to significant transformation. Thus we study social class differences in development, ethnic differences, rural-urban differences—or, at the next level down, children from one- versus two-parent homes, large versus small families—as if the nature of these structures, and their developmental consequences, were eternally fixed and unalterable except, perhaps, by violent revolution. We are loath to experiment with new social forms as contexts for realizing human potential. "After all," we say, "you can't change human nature." This precept underlies our national stance on social policy and much of our science of human development as well.

Research on macrosystem change requires a shift in the nature of the contrasts to be employed in experiments. It is one thing to compare the effects on development of systems or system elements already present within the culture; it is quite another to introduce experimental modifications that represent a restructuring of established institutional forms and values.

The last, and most demanding, of the basic definitions outlining the nature and scope of research on the ecology of human development identifies a strategy of choice for scientific work in this sphere.

DEFINITION 11
A transforming experiment involves the systematic alteration and restructuring of existing ecological systems in ways that challenge the forms of social organization, belief systems, and lifestyles prevailing in a particular culture or subculture.

A transforming experiment systematically alters some aspect of a macrosystem. The alteration may be effected at any level of the ecological environment from the micro- to the exosystem by eliminating, modifying, or adding elements and interconnections.

A general principle pervades all the basic concepts for an experimental ecology of human development. The principle is stated as the first of a series of propositions describing the distinctive characteristics of research models appropriate for investigating development-in-context.

PROPOSITION A
In ecological research, the properties of the person and of the environment, the structure of environmental settings, and the processes taking place within and between them must be viewed as interdependent and analyzed in systems terms.

The specification of these interdependencies constitutes a major task of the proposed approach. The rest of this volume represents a beginning effort in this direction. In the chapters that follow, I outline in greater detail the distinctive properties of an ecological model, in terms of both theory and research design, that are appropriate for analyzing developmental contexts and processes at each of the four environmental levels. At each level, I have provided one or more concrete examples of investigations—actual when available, hypothetical when not—to illustrate these distinctive properties, by either demonstration or default.

For reasons already indicated, well-designed ecological experiments are, as yet, not easy to find. I have therefore had to invent some examples where they did not exist. Moreover, in many instances there was a dearth not only of relevant research but also of relevant research ideas. Accordingly, the chapters that follow contain even more proposed hypotheses than proposed investigations.

Since the proposed hypotheses have never been tested, at least in the form and context in which they are presented, there is typically no empirical evidence bearing directly on their validity. Nevertheless, in selecting research examples for presentation, I have endeavored to pick those that illustrate at least the promise of the posited relationships. Such evidence, however, will be mostly circumstantial and never compelling or complete. For the present, therefore, the hypotheses can be judged and justified only on theoretical grounds. The ultimate test of empirical investigation still lies ahead.

When and if the test comes, the hypotheses may prove invalid, but that is an outcome that, in science, is neither uncommon nor unrespectable. The proposed investigations, however, may suffer a less honorable fate. Since they are research ideas that have never been tried out, what German psychologists have called Gedanken experiments, the effort to implement them may well reveal fatal flaws in conception, design, or feasibility. But I hope that at the very least they will point the way to fruitful scientific discoveries by future investigators.

PART TWO

Elements
of the Setting

3.

The Nature and Function of Molar Activities

I begin my consideration of the elements of the microsystem with a discussion of molar activities because these constitute the principal and most immediate manifestation both of the development of the individual and of the most powerful environmental forces that instigate and influence that development—the actions of other people. To be more explicit: molar activities as exhibited by the developing person serve as indicators of the degree and nature of psychological growth; as exhibited by others present in the situation they constitute the principal vehicle for the *direct* influence of the environment on the developing person.

All molar activities are forms of behavior, but not all behaviors are forms of molar activity. The reason for making the distinction lies in the belief that not all behaviors are equally significant as manifestations of or influences on development. Many are so short-lived as to have minimal import; these are referred to as *molecular* behaviors. Others are more long-lasting but, because they lack meaning to the participants in the setting, have only negligible impact. The definition of molar activity thus emphasizes both some persistence through time and some salience in the phenomenological field of the developing person and of others present in the setting.

DEFINITION 12
A molar activity is an ongoing behavior possessing a momentum of its own and perceived as having meaning or intent by the participants in the setting.

The terms *molar* and *ongoing* are used to emphasize that an activity is more than a momentary event, such as a movement or an utterance; rather, it is a *continuing process* that entails more

45

than a beginning or an end. A molar activity is distinguished from an *act*, which is perceived as instantaneous and hence molecular in character. Examples of acts are a smile, a knock on the door, a single question, or an answer. The following are molar activities: building a tower of blocks, digging a ditch, reading a book, or carrying on a telephone conversation.

A second, even more distinctive property of molar activities and one that was particularly emphasized by Lewin and his students, is the fact that they are characterized by a *momentum* of their own, a tension system (Lewin, 1935) that makes for persistence through time and resistance to interruption until the activity is completed (Ovsiankina, 1928). For the most part, this momentum is produced by the existence of intent (Birenbaum, 1930)—the desire to do what one is doing either for its own sake or as a means to an end. The presence of intent creates a motive for closure, which in turn leads to perseverance and resistance to interruption. Some molar activities are not characterized by intent, at least in the form of a conscious goal (for example, sleeping, daydreaming, or running aimlessly around the room), but in these cases intention is conspicuous by its absence. The question of perceived aim is thus always relevant for defining an activity, if only by default.

Putting the issue another way, activities vary in the degree and complexity of the purposes that animate them. This variation is reflected along two additional dimensions that are completely phenomenological in character, meaning that they are defined according to how they are perceived by the actor. The first of these subjective domains is *time perspective*, determined by whether the actor perceives the activity as taking place only in the immediate present as she engages in it or as part of a larger temporal trajectory, transcending the bounds of ongoing action, reaching back into the past or forward into the future. This last component, of anticipation, often intersects with the second phenomenological domain: the extent to which the activity is consciously perceived as having an explicit *goal structure*, whether the path to the goal is perceived as direct, involving a single course of action (such as climbing to reach a desired object) or as involving a sequence of steps or subgoals, consisting of a series of preplanned stages (exemplified by organizing a beach hike with younger siblings to look for shells from which to make mother a necklace for next Christmas).

Activities with a complex goal structure typically involve an extended time perspective as well, but the converse is not necessarily true. The goal structure may be quite simple, consisting of only a single course of action, but entail a long delay of gratification, as in saving up money in a piggy bank to buy a toy.

Another dimension along which molar activities can vary in complexity extends well beyond the parameters of time perspective and goal structure. Activities differ in the *extent to which they invoke objects, people, and events not actually present in the immediate setting.* Such invocation may be accomplished through conversation, story telling, fantasy, pictorial representation, or a variety of other media. To the extent that activities refer to events occurring in other places at other times, they reflect an expansion of the actor's phenomenological world beyond the immediate situation. Thus it is possible to speak of an "ecology of mental life" with a potential structure isomorphic with that of the ecological environment. If a person in a given setting speaks about her own activities in some other setting, either in the past or in the future, she is exhibiting the ability to create a "mental mesosystem." Television brings into the daily experience of children violent events in other places that then find violent expression in the youngster's everyday activities, thus adding an exo- and, perhaps, even more tragically, an entire macrosystem to the child's phenomenological world.

Even when a person's activities are restricted to experiences in and of the immediate setting, they can take on a high order of complexity through the introduction of another element of the microsystem, *relations with other people.* Although many molar activities can be carried out in solitude, some necessarily involve interactions with other persons. Children in particular spend much time in joint activities with adults or age-mates. In the beginning these tend to be dyadic, involving only one other person at a time. But soon the child is able to be aware of and to deal with two or more persons simultaneously, thus maintaining and eventually even creating what are later defined in the ecological schema as $N + 2$ systems.

The fact that the child becomes able to establish complex interpersonal relationships on her own reflects an important principle in the ecology of human development: as the child's phenomenological field expands to include ever wider and more differentiated aspects of the ecological environment, she becomes capable not only of participating actively in that environment but also of modifying and adding to its existing structure and content.

Finally, as the child develops, she becomes capable of *carrying on more than one molar activity at a time.* Although there is no research bearing on the question, it is possible that children acquire and perfect this skill through contact with parents, especially mothers, who, usually by necessity, become proficient in dealing with their children while continuing to engage in one or more other essential activities.

The emerging molar activities of the child reflect the evolving scope and complexity of the perceived ecological environment, both within and beyond the immediate setting, as well as the child's growing capacity to deal with and alter this environment in accord with his needs and desires. Molar activities are important in yet another respect: when exhibited by others present in the setting, they constitute the main source for *direct* effects of the immediate environment on psychological growth. It follows from the preceding exposition that the development of the child is a function of the scope and complexity of the molar activities engaged in by others that become part of the child's psychological field either by involving her in joint participation or by attracting her attention.

In keeping with the Lewinian precedent, no mention has been made thus far of the substantive nature of molar activities, as distinguished from their structural properties. The reason for not specifying subject matter in advance has already been stated: the question is an empirical one that can be answered only after relevant data have been obtained. It is here that we encounter a major obstacle to more detailed specification of a theoretical framework for an ecology of human development. Researchers as yet know very little about the molar activities of children and their caretakers in the actual settings in which people live out their lives. Laboratory studies have yielded voluminous data about molecular acts, but information about larger behavioral units in natural environments is far more sparse. A notable exception is the painstaking research of Barker, Wright, and their colleagues (Barker and Wright, 1954; Barker and Gump, 1964; Wright, 1967; Barker and Schoggen, 1973) on the "psychological ecology" of childhood. My own theory builds on their work, but departs from it in a number of important ways. First, although similarly emphasizing the importance of studying behavior at a molar level, Barker and his associates tend to concentrate on the *process* of interaction rather than its *content*. Thus most of the analyses involve such variables as dominance, nurturance, compliance, and avoidance rather than categories dealing with the substance of the activity in the course of which these patterns of relationships were displayed. Second, the focus of attention is on the behavior of individuals taken one at a time; for example, the researcher analyzes the behavior of children, or of caretakers, but not of child and caretaker as a dyadic unit. To state the same point in another way: the behavior of the individual is classified without regard to its relation to the behavior of other persons present in the situation. In short, activities are not viewed in their interpersonal context. Third, consistent with this orientation, the setting is conceived in purely behavioral terms without reference to social structure either in the immediate or the more remote environment. Finally, there is no attempt to examine molar activity from a developmental perspective, to view its complexity and content as reflecting the level of the person's psychological growth. In sum, neither the properties of the person nor of the environment are conceptualized in systems terms.

In the absence of concepts, methods, and data bearing on the content and interpersonal structure of molar activities exhibited in settings of everyday life by persons at varying stages of development, my colleagues and I have undertaken to make a beginning in this threefold task (Nerlove et al., 1978). As a point of departure, we chose to investigate the ongoing behaviors of three- to five-year-old children and their caretakers both at home and in preschool settings, including nursery schools and day care centers. We defined as our initial objective the development of a taxonomy of molar activities in terms of their content, complexity, and interpersonal structure. Two general methodological approaches were employed. In the first an observer was requested to focus on the activities of a particular child and the people around him, and to describe in his own words what the child was doing and what the people around him were doing. The field workers engaged in this task were familiar with both kinds of settings under investigation and came from cultural backgrounds similar to those of the persons being studied. The observers were instructed to describe activities from the perspective of the participants in the setting. The second strategy involved asking the child's principal caretaker, usually the mother, to provide a similar description for one segment of a day—a morning, afternoon, or evening, each including a meal.

Both sets of protocols were subjected to content analysis to identify the categories spontaneously employed for describing those behaviors, of both children and their caretakers, that were judged subsequently by independent coders to meet our criteria for molar activity. Each activity was classified in four general spheres: content; "psychological momentum," as indicated by initiative, level of concentration, resistance to distraction, resumption after interruption, and so on; complexity of activity structure, as manifested in the number of molar activities carried on simultaneously, extended time perspective, and the presence of sequential subgoals; and the complexity of the perceived ecological field, as reflected in the person's participation in interpersonal systems (dyad, triad, and so on), reference to events in other settings, and the modification or expansion of the life space through fantasy or actual reconstruction of the objective environment.

Satisfactory interjudge reliability ($r = .70$ to $.80$) has been obtained in the coding of parallel running records prepared by pairs of observers independently describing the same events over fifteen-minute periods. The taxonomy will be cross-validated in an ongoing comparative study of the ecology of children and families in five modern industrialized societies (Bronfenbrenner and Cochran, 1976).

The statistical analysis of pilot study data is still under way and, beyond reliability figures, no systematic findings are as yet available. It is instructive, however, to examine the content of the molar activities that have been reported for the American sample. Categories derived from the content analysis of observations in about twenty-five families, and interviews with more than one hundred mothers, fall into the following general domains. At the more passive extreme the first domain, entitled "nonengagement," consists of such pursuits as sleeping, resting, drifting (wandering aimlessly around); the most focused behavior in this domain is waiting. A second sphere contains activities that involve paying attention to people or ongoing events without active participation. Other areas are characterized by enduring emotional states, nonfantasy and fantasy play, games, musical activity, responsibilities and work, educational processes, and activities with a predominantly social purpose.

Each activity is also analyzed for complexity as reflected in simultaneity with other ongoing behaviors, time perspective, goal structure, extent of involvement in an interpersonal system (dyad, triad, and so on), and reference in conversation, fantasy play, or symbolic representation to events, objects, or people not present in the immediate situation.

It may also be instructive to consider the potential significance of the several types of molar activities exhibited by the children in the

American pilot study. Given a theoretical perspective emphasizing the importance of motivational momentum and complexity in the structure of goals and interpersonal systems, activities of nonengagement (for instance, sleeping, resting, daydreaming, wandering aimlessly about, being restlessly hyperactive) are presumed to constitute the lower bound of the developmental continuum. Children observed as spending much of their time in such activities are viewed as less advanced in their psychological growth. At the same time, in keeping with a dynamic concept of the human organism, their preoccupation with these pursuits is seen as an effort to establish or find conditions in which they could function more effectively. The same interpretation applies with even greater force to emotional activities, both negative (such as protracted crying, expressions of anger, or fighting) and positive (joyful states, continuing expressions of affection or approval). Again, these are regarded as attempts either to alter circumstances that impair the capacity to function or to perpetuate and enhance siuations that facilitate developmental processes. The validity of these assumptions must be determined empirically by investigation of the behavioral correlates and consequences of activities of nonengagement and their longer-range sequelae in subsequent patterns of molar activity in other settings.

The domain of attending—paying heed to other people and events—is developmentally significant in constituting the necessary condition for observational learning. Whether such learning in fact occurs can, again, be determined by investigating whether the child subsequently tries to carry out activities he has seen others conduct.

The relevance of educational and musical activities for learning and development is self-evident. But the remaining domains of nonfantasy and fantasy play, games, responsibilities and work, and social activities merit discussion, particularly since they are not accorded high priority in American research on socialization processes and outcomes, or—for that matter—in actual socialization activities taking place in American society.

This neglect is particularly marked for play, fantasy, and games. Although the importance of such activities to developmental processes has been stressed in the theoretical writings and clinical observations of Piaget (1962), the translation of these ideas into research and practice has been minimal, at least in the United States. In a number of other societies, however, play, fantasy, and games are topics of extended scientific study, and the results serve as the basis of recommended practice in homes, preschools, and school curricula. The Soviet Union is a case in point. The research emphasis stems from the theories of Vygotsky and his disciples (Elkonin, 1978; Leontiev, 1964; Vygotsky, 1962, 1978; Zaporozhets and Elkonin, 1971), who view play, fantasy, and games as important activities for cognitive, motivational, and social development. Proceeding from this theoretical base, Soviet pedagogues have incorporated many play activities, both imaginary and real, into the preschool and elementary curriculum (Venger, 1973; Zaporozhets and Elkonin, 1971; Zaporozhets and Markova, 1976; Zhukovskaya, 1976). As the children grow older, increased importance is accorded to the educational benefits of what the Russians call *role-vaya igra,* role-playing games, in which children take roles that are common in adult society, for instance, store clerk, customer, nurse, patient, and so on. A fuller description of such activities appears elsewhere (Bronfenbrenner, 1970a).

Soviet educators use play, fantasy, and games primarily to develop what they refer to as "communist morality." From an American perspective, the Russian outcome would be viewed as representing a remarkably high level of social conformity and submission to authority. These effects are documented in a series of experiments on reactions to social pressure on the part of Soviet school children compared with age-mates from the United States and other Western societies (Bronfenbrenner, 1967, 1970b; Garbarino and Bronfenbrenner, 1976; Kav-Venaki et al., 1976; Shouval et al., 1975). There is reason to believe that play, fantasy, and games can be just as effectively utilized to develop initiative, independence, and equalitarianism. Indeed such activities probably function precisely in this fashion in contemporary American settings both within and outside school. The relevant research has yet to be carried out, and will require an appropriate taxonomy of activities that extends to children of elementary school age and beyond. One can anticipate, however, that various aspects of play, fantasy, and games will relate not only to the development of conformity versus autonomy but also to the evolution of particular forms of cognitive function. It is noteworthy in this regard that, in the course of pilot-testing our activities code, we observed the most complex cognitive operations in the realm of fantasy play.

The relevance of social and work-related activities to human development can be expressed in two statements for which no research documentation as yet exists (a fact that, paradoxically, can be interpreted as reflecting the validity of the statements themselves). First, in the United States it is now possible for a person eighteen years of age to graduate from high school without ever having had to do a piece of work on which somebody else truly depended. If the young person goes on to college, the experience is postponed for another four years. If he goes on to graduate school, some might say the experience is postponed forever.

The second statement points to what may be an even more destructive outcome in the long run. In the United States, it is now possible for a person eighteen years of age, female as well as male, to graduate from high school, college, or university without ever having cared for, or even held, a baby; without ever having looked after someone who was old, ill, or lonely; or without ever having comforted or assisted another human being who really needed help. Again, the psychological consequences of such a deprivation of human experience are as yet unknown. But the possible social implications are obvious, for—sooner or later, and usually sooner—all of us suffer illness, loneliness, and the need for help, comfort, or companionship. No society can long sustain itself unless its members have learned the sensitivities, motivations, and skills involved in assisting and caring for other human beings.

Yet the school, which is the setting carrying primary responsibility for preparing young people for effective participation in adult life, does not, at least in American society,[1] give high priority to providing opportunities in which such learning could take place. This would not be impossible to achieve. For some years I have been advocating the introduction in our schools, from the earliest grades onward, of what I have called a *curriculum for caring* (Bronfenbrenner, 1974b, 1974c, 1978b). The purpose of such a curriculum would be not to learn *about* caring, but to engage *in* it: children would be asked to take responsibility for spending time with and caring for others—old people, younger children, the sick, and the lonely. It would be essential that such activities be carried out under firm supervision, and this supervision could not be provided by already overburdened teachers. Instead, the supervisors should be drawn from persons in the community who have experience in caring—parents, senior citizens, volunteer workers, and others who understand the needs of those requiring attention and the demands on those who would give it. Obviously such caring activities cannot be restricted to the school—they will have to be carried on in the outside community. It would be desirable to locate caring institutions, such as day care centers, adjacent to or even within the school. But it would be even more important for the young caregivers to come to know the circumstances in which their charges live and the people in their lives. For example, an older child taking responsibility for a younger one might come to know the latter's family and become acquainted with his neighborhood by escorting him home from school. In this way, the older children, as well as the adults involved in the program, would learn to know at first hand the living conditions of the people in their community.

My purpose in describing the proposed program here is not to advocate its adoption but to illustrate ecological concepts, their concrete implications for developmental research, and the essential interplay between issues of public policy and basic science in the study of development-in-context. Viewing the suggested project first from a social policy perspective, it is clear that before any such curriculum for caring is introduced on a broad scale, it should be tested experimentally and its putative effects, along with possible unintended consequences, evaluated. But once it becomes a research enterprise, an effort of this kind also constitutes an excellent example of what I have called a transforming experiment, since it calls into question and alters in a substantial way a prevailing pattern in the American macrosystem, the current "blueprint" for what a school curriculum should and should not contain. Indeed as a scientific undertaking, the proposed program entails changes at all four levels of the ecological environment. Thus it reaches beyond the microsystem of the classroom to invoke new interconnections among home, school, and neighborhood at the level of the mesosystem. To the extent that adults from the community who become directly involved with the program are influenced to introduce changes in other settings in which they participate (for instance, committees, offices, and organizations), the curriculum may also have exosystem effects. Within the most immediate environmental sphere, the suggested project involves alteration not only in molar activities but also in the microsystem elements of role and interpersonal structure. It is in fact by introducing changes in the traditional role expectations for pupils and children that new activities involving new patterns of social interactions are set in motion. As we shall discover (chapter 5), the creation and allocation of roles is an especially powerful strategy for influencing the course of human development.

Even if the proposed curriculum did not bring about significant change in the prevailing conception of what schools are or should be trying to accomplish, it would still be important both for science and public policy to document the kind of molar activities that are now occurring in our classrooms. For the availability of such information, whether for schools or other human habitats (homes, day care centers, playgrounds, peer group hangouts, places of work, retirement homes, and so on), would permit assessment of both the developmental status of the person and the power of the "activity milieu" to stimulate or stifle psychological growth.

Molar activities thus have manifold functions in regard to human development, since they can serve equally, and sometimes simultaneously, as cause, context, and consequence of psychological growth. But for purposes of research, it is necessary to keep these functions separate; in particular, neither cause nor context should be confounded with outcome. In accord with this methodological principle, the conclusion concerning the significance of molar activities for development is stated in two parts—the first, a proposition dealing with molar activities as developmental outcomes, and the second, a hypothesis setting forth the function of the activity milieu as a context and potential influence on developmental processes.

PROPOSITION B
The developmental status of the individual is reflected in the substantive variety and structural complexity of the molar activities which she initiates and maintains in the absence of instigation or direction by others.

Substantive variety refers to the range in content of these activities. Structural complexity is manifested in the evolving scope and differentiation of the developing person's perceived ecological environment, both within and beyond the immediate setting, as well as in her growing capacity to deal with and alter that environment in accord with her own needs and desires.

HYPOTHESIS 1
The development of the person is a function of the substantive variety and structural complexity of the molar activities engaged in by others who become part of the person's psychological field either by involving her in joint participation or by attracting her attention.

As the two foregoing statements imply, the person's perceptions of and interaction with others, in both the immediate and the more remote environment, are especially salient both as influences on and manifestations of development. The emerging structure and content of these relations, and their developmental implications, are thus of particular interest to us.

4.

Interpersonal Structures as Contexts of Human Development

We begin with a definition.

DEFINITION 13
A relation obtains whenever one person in a setting pays attention to or participates in the activities of another.

The presence of a relation in both directions establishes the minimal and defining condition for the existence of a *dyad*: a dyad is formed whenever two persons pay attention to or participate in one another's activities.

The dyad is important for development in two respects. First, it constitutes a critical context for development in its own right. Second, it serves as the basic building block of the microsystem, making possible the formation of larger interpersonal structures—triads, tetrads, and so on. In terms of its potential for furthering psychological growth, there are three different functional forms that a dyad may take.

1. An *observational dyad* occurs when one member is paying close and sustained attention to the activity of the other, who, in turn, at least acknowledges the interest being shown. For example, a child watches closely as a parent prepares a meal and makes occasional comments to the child. This type of dyad obviously meets the minimal condition necessary for observational learning, but stipulates an additional interpersonal requirement: not only must the activity of the other person actually be a focus of attention, but also that person must make some overt response to the attention being shown. Once an observational dyad is in existence, it readily evolves into the next, more active dyadic form.

2. A *joint activity dyad* is one in which the two participants per-

56

ceive themselves as doing something together. This does not mean that they are doing the same thing. On the contrary, the activities each engages in usually tend to be somewhat different, but complementary—part of an integrated pattern. For example, parent and child may be looking at a picture book; the mother tells the story, while the child names objects in response to her questions. A joint activity dyad presents especially favorable conditions not only for learning in the course of the common activity but also for increasing motivation to pursue and perfect the activity when the participants are no longer together.

The developmental power of a joint activity dyad derives from the fact that it enhances, and thereby exhibits in more marked degree, certain properties that are characteristic of all dyads.

Reciprocity. In any dyadic relation, and especially in the course of joint activity, what A does influences B and vice versa. As a result, one member has to coordinate his activities with those of the other. For a young child, the necessity of such coordination not only fosters the acquisition of interactive skills, but also stimulates the evolution of a concept of interdependence, an important step in cognitive development.

Furthermore reciprocity, with its concomitant mutual feedback, generates a momentum of its own that motivates the participants not only to persevere but to engage in progressively more complex patterns of interaction, as in a ping-pong game in which the exchanges tend to become more rapid and intricate as the game proceeds. The result is often an acceleration in pace and an increase in complexity of learning processes. The momentum developed in the course of reciprocal interaction also tends to carry over to other times and places: the person is likely to resume his or the other person's "side" of the joint activity in other settings in the future, either with others or alone. It is in this way that dyadic interaction, especially in the course of joint activity, produces its most powerful developmental effects.

Balance of power. Even though dyadic processes are reciprocal, one participant may be more influential than the other. For example, in a tennis game, one player, during a long volley, drives the other into a corner. The extent to which, in a dyadic relation, A dominates B is referred to as balance of power. This dyadic dimension is important for development in several respects. For a young child, participation in dyadic interaction provides the opportunity for learning both to conceptualize and to cope with differential power relations. Such learning contributes simultaneously to cognitive and social development, since power relations characterize physical as well as social phenomena encountered by the growing person in a variety of ecological settings throughout the life span.

Balance of power is significant in yet another, more dynamic respect, since there is evidence to suggest that the optimal situation for learning and development is one in which the balance of power gradually shifts in favor of the developing person, in other words, when the latter is given increasing opportunity to exercise control over the situation.

Joint activity dyads are especially well suited to this developmental process. They stimulate the child to conceptualize and cope with power relations. At the same time, they provide an ideal opportunity for effecting a gradual transfer of power. Indeed this transfer often takes place "spontaneously" as a function of the active character of the developing person in relation to the environment.

Affective relation. As participants engage in dyadic interaction, they are likely to develop more pronounced feelings toward one another. These feelings may be mutually positive, negative, ambivalent, or asymmetrical (as when A likes B, but B dislikes A). Such affective relations tend to become more differentiated and pronounced in the course of joint activity. To the extent that they are positive and reciprocal to begin with and become more so as interaction proceeds, they are likely to enhance the pace and the probability of occurrence of developmental processes. They also facilitate the formation of the third type of two-person system, a primary dyad.

3. A *primary dyad* is one that continues to exist phenomenologically for both participants even when they are not together. The two members appear in each other's thoughts, are the objects of strong emotional feelings, and continue to influence one another's behavior even when apart. For example, a parent and child, or two friends, miss each other when they are not together, imagine what they might be doing, what the other one might say, and so on. Such dyads are viewed as exerting a powerful force in motivating learning and steering the course of development, both in the presence and absence of the other person. Thus a child is more likely to acquire skills, knowledge, and values from a person with whom a primary dyad has been established than from one who exists for that child only when both are actually present in the same setting.

Although each has its distinctive properties, the three dyadic forms are not mutually exclusive; that is, they can occur simultaneously as well as separately. A mother and her preschool child's reading a book together is obviously a joint activity taking place in the context of a primary dyad. But if the child's part is mainly one of listening attentively as the mother reads aloud, the dyad is clearly also an observational one. As might be expected, such combined structures have a more powerful developmental impact than dyads confined to a single type. This point will be taken into account in the consideration of specific dyadic hypotheses and methods for their investigation.

The dyadic properties and principles I have outlined may be summarized in the form of a series of hypotheses describing the presumed impact of various types of dyadic structure on developmental processes. I begin by calling attention to an evolutionary process at the level of the dyad itself. The first two hypotheses postulate that dyads can undergo a course of development just as individuals do.

HYPOTHESIS 2
Once two persons begin to pay attention to one another's activities, they are more likely to become jointly engaged in those activities. Hence observational dyads tend to become transformed into joint activity dyads.

HYPOTHESIS 3
Once two persons participate in a joint activity, they are likely to develop more differentiated and enduring feelings toward one another. Hence joint activity dyads tend to become transformed into primary dyads.

The next hypothesis specifies the dyadic properties conducive to development.

HYPOTHESIS 4
The developmental impact of a dyad increases as a direct function of the level of reciprocity, mutuality of positive feeling, and a gradual shift of balance of power in favor of the developing person.

The hypotheses that follow deal with the joint effects produced when different kinds of dyads occur simultaneously.

HYPOTHESIS 5
Observational learning is facilitated when the observer and the person being observed regard themselves as doing something together. Thus the developmental impact of an observational dyad

tends to be greater when it takes place in the context of a joint activity dyad (a child is more likely to learn from watching a parent cook a meal when the activity is structured so that the two are acting together).

HYPOTHESIS 6
The developmental impact of both observational learning and joint activity will be enhanced if either takes place in the context of a primary dyad characterized by mutuality of positive feeling (one learns more from a teacher with whom one has a close relationship). Conversely, mutual antagonism occurring in the context of a primary dyad is especially disruptive of joint activity and interferes with observational learning.

Finally, if all these considerations are taken into account, one can stipulate the optimal conditions for learning and development in a dyadic relationship.

HYPOTHESIS 7
Learning and development are facilitated by the participation of the developing person in progressively more complex patterns of reciprocal activity with someone with whom that person has developed a strong and enduring emotional attachment and when the balance of power gradually shifts in favor of the developing person.

The question may be raised whether a positive relationship between the members of a dyad is essential as long as the participants continue to engage in progressively more complex patterns of reciprocal activity. The question assumes that the second condition is independent of the first. I shall cite evidence from research for the debilitating impact of antagonism between the participants on the functioning of the dyad as a developmental system.

Because of their focal role in the ecology of human development, it is convenient to have a single term for dyads that meet the optimal conditions, stipulated in hypotheses 4 through 7, of reciprocity, progressively increasing complexity, mutuality of positive feeling, and gradual shift in balance of power. Accordingly, I will refer to two-person systems exhibiting these properties as *developmental dyads*.

Although studies dealing with dyads are fairly common in the literature of both social and developmental psychology, few of these investigations bear directly on issues of development, once again for the reason that they are limited to a single setting at a single point in time and hence do not meet the criterion of developmental validity. Even rarer are researches that provide evidence for or against the dyadic hypotheses. One series of experiments and follow-up studies, however, despite a glaring omission in the data, dramatically documents the motivating power and long-range developmental effect of the dyad as a context for development.

The thesis that behavior in dyads is generally reciprocal is widely accepted in theory, but it is often disregarded in research practice. The failure to take two-way processes into account reflects the inertia of the traditional laboratory model with its classical participants —an experimenter, identified cryptically as E, and another person equally informatively described as S, the subject. The term *subject* is apt, for with few exceptions the process operating between E and S is viewed as unidirectional; the experimenter presents the stimulus, and the subject gives the response. Of course in theory the influence can occur in both directions, but once the researcher puts on the white coat of scientific invisibility, she tends to focus solely on the behavior of the experimental subject, even when someone besides the experimenter is an active participant in the setting.

A case in point is the work of Klaus, Kennell, and their colleagues at the Case Western Reserve School of Medicine (Hales, 1977; Hales, Kennell, and Susa, 1976; Kennell, Traus, and Klaus, 1975; Kennell et al., 1974; Klaus and Kennell, 1976; Klaus et al., 1970, 1972; Ringler, 1977; Ringler et al., 1975). The investigators took as their point of departure observations on animals revealing complex, species-specific patterns of mother-neonate interaction immediately after delivery (Rheingold, 1963). Their aim was to explore this phenomenon in humans. Noting that still-prevailing hospital practices resulted in minimal opportunities for contact between mother and newborn, the researchers modified the established procedures to permit mothers to have their naked infants with them for about an hour shortly after delivery and for several hours daily thereafter. To avoid chilling, a heat panel was provided over the mothers' beds. Randomly assigned control groups experienced the type of "contact with their babies that is routine in American hospitals (a glance at their baby shortly after birth, a short visit six to twelve hours after birth for identification purposes, and then 20- to 30-minute visits for feeding every four hours during the day)" (Kennell et al., 1974, p. 173). To insure comparability a heat panel was installed over the control mothers' beds as well. Neither group knew that the other was being treated differently.

The reported results of these experiments strain the credulity of the reader. In the initial experiment (Klaus et al., 1970), all mothers of full-term infants in the extended exposure group exhibited "an orderly progression of behavior": "The mothers started with fingertip touch on the infants' extremities and proceeded in 4 to 8 minutes to massaging, encompassing palm contact on the trunk ... Mothers of normal premature infants permitted to touch them in the first 3 to 5 days of life followed a similar sequence, but at a much slower rate" (p. 187). The mothers of full-term babies in the experimental treatment also "showed a remarkable increase in the time spent in the 'en face' position in only 4 to 5 minutes" (p. 190).

In a second study (Klaus et al., 1972) with a new sample, fourteen "extended-contact" mother-infant pairs and an equal number of randomly assigned controls, well matched on developmental and family background factors, were compared when their children were one month old. All the mothers were primiparous, with healthy, full-term infants. In this and other follow-up studies, none of the observers knew to which group the subjects belonged. During a hospital examination one month after birth, the mothers in the extended-contact group significantly more often stood and watched beside the examination table and soothed their babies when they cried. They also showed greater fondling and eye-to-eye contact while feeding their babies and, in an interview, expressed greater willingness to pick up their infants when they fussed and more reluctance and anxiety about leaving the baby in someone else's care. Moreover, these differences were still in evidence when the infants were reexamined at one year of age (Kennell et al., 1974). The mothers in the extended-contact group reported missing the baby more when separated from it; during the physical examination, they were again more likely to stand by the tableside and assist the physician, to soothe the infant when it cried, and to kiss their babies.

In a subsequent follow-up study (Ringler et al., 1975), when the infants were two years old, the mother's conversation with the child was observed and recorded during a free play period in a setting containing toys and books. "Speech patterns of the mothers revealed that those who had been given extra contact with their infants during the neonatal period used significantly more questions, adjectives, words per proposition, and fewer commands and content words than did the control mothers" (p. 141).

The most recent experiment in the series (Hales, Kennell, and Susa, 1976) not only provides a much-needed replication of the initial studies with a larger sample ($N = 60$) but does so in a different cultural context and with a more rigorous experimental design that permits resolving the issue of whether there exists a critical period of susceptibility to extended contact between mother and infant. Although the original investigators spoke of "a special attachment period for an adult woman" (Klaus et al., 1972, p. 463), they acknowledged that their data left open the question of timing: was it a matter of the first few hours after birth, or extended contact over the next several days? In the latest experiment carried out at Roosevelt Hospital in Guatemala, Hales and her associates clarified this issue by introducing two early-contact groups, one limited to forty-five minutes immediately after delivery and the second for an equal interval but beginning twelve hours after the infant's birth. The results were unequivocal. Only the mothers in the immediate contact group were affected:

Mothers who had contact with their neonates immediately after birth showed significantly more affectionate behavior ("en face," looking at the baby, talking to the baby, fondling, kissing, smiling at the infant) when compared to the mothers in the delayed and control groups ... No significant differences were noted between the delayed and control groups. This study indicates that the maternal sensitive period is less than twelve hours in length, suggests the importance of skin to skin contact and compels reconsideration of hospital practices that even briefly separate mother and infant. (Hales, Kennell, and Susa, 1976, p. 1)

From an ecological perspective, even more remarkable than the dramatic results reported in this series of experiments are the data they omit. In none of the papers cited is there a single word about the behavior of the infant in the mother-infant dyad, and all the experimental effects are attributed entirely to the mother. Thus the investigators refer repeatedly to a "maternal sensitive period" (Klaus et al., 1972, p. 463) or a special attachment or sensitivity period existing "in the human mother" (Kennell et al., 1974, p. 173; Kennell, Trause, and Klaus, 1975, p. 87). Given the dyadic property of reciprocity, the question naturally arises whether the distinctive behavior of the mothers in the experimental group during the initial early contact, subsequent extended exposure, and later follow-up might not have occurred, at least in part, as a response to a sequence of activities *initiated by the developing infant* and reciprocated by the mother in a progressively evolving pattern of social interaction. The possibility remains unexplored. In keeping with the classic experimental model, the focus of scientific attention in these studies was limited to the subjects of the research, who in this instance were not the children but the mothers. The omission is all the more striking given the fact that not only were the infants always present in the research situation, but all the mother's behavior being observed was directed toward them.

To be sure, in the most recent reports (Kennell, Trause, and Klaus, 1975; Ringler, 1977) follow-up data are reported on the children's developmental status at age five as related to the mother's behavior toward the child at younger ages. Although the reports still do not provide any information about the behavior of the infant toward the mother at the earlier period, the results nevertheless merit serious consideration. Ringler found that, in comparison with controls, "the five-year-olds of the early contact mothers had significantly higher IQ's, understood language as measured by a receptive language test significantly better and comprehended significantly more phrases with two critical elements" (p. 5). The IQ difference was approximately seven points. Furthermore, there were significant correlations between measures of the complexity of speech patterns employed by mothers toward their infants when the latter were two years of age and indexes of the child's level of language comprehension and performance at age five. The Pearson product-moment coefficients ranged from .72 to .75. A significant correlation of .71 was found between the child's IQ at age five and "the amount of time women spent looking at their babies during the filmed feeding at one month of age" (Kennell, Trause, and Klaus, 1975, p. 93).

These latest findings are urgently in need of replication, especially in view of the small number of subjects. Nevertheless, despite the lamentable absence of data on the infant's side of the dyadic equation, this series of experiments presents persuasive evidence for the scientific utility and promise of the concepts and hypotheses presented above. Thus the studies provide a glimpse—albeit tantalizingly one-sided—of the process through which the joint activity of mother and newborn leads to the formation of a primary dyad, which in turn sets the pace and steers the course of future development. Because one sees only the mother's part of the interaction, it is impossible to assess the level of reciprocity, the degree of mutual positive feeling, or the shift in balance of power from mother to infant (or perhaps from infant to mother?). As in listening to one side of an animated telephone conversation, one may sense the back and forth movement, the response in kind, and the rise and fall of pressure coming from the other end of the line. But scientists have yet to record the two ends of the "conversation" simultaneously and especially to trace the resultant trajectory of development for *both* parties.

An important theoretical insight was thus ironically provided by the one-sided focus of the Western Reserve studies. By document- ing the evolution of the mother's behavioral and emotional involve- ment with the infant, rather than the reverse, the investigators showed that in the course of dyadic interaction the mother is living through a developmental experience no less profound or consequen- tial than that experienced by her offspring. In keeping with our conception of and criteria for developmental change, the mother does indeed manifest a progressively more extended and differen- tiated view of a newly prominent aspect of her environment (that is, the arrival of her child) and becomes motivated and able to undertake new activities in dealing with the environment that are of a high order of complexity in form and content. And, what is most critical for establishing that development has in fact taken place, these newly developing perceptions and activities clearly have their sequelae in other places and at other times, in this instance as much as five years later.

Since it is safe to assume that the child, too, has experienced psy- chological growth during this period, we arrive at a key proposition regarding the developmental properties of a dyad.

PROPOSITION C
If one member of a dyad undergoes developmental change, the other is also likely to do so.

The basic principle underlying this proposition is, of course, in no way new. With respect to mother-infant interaction, it received its definitive statement a decade ago in Harriet Rheingold's classic paper entitled "The social and socializing infant" (1969a). As far as research practice is concerned, however, the principle has been mainly ignored. Indeed the Western Reserve experiments, even though they do not view the dyad as a reciprocal system, represent a step beyond typical experimental studies, which, being limited to a single setting and point in time, cannot provide evidence for the occurrence of enduring developmental effects as distinguished from temporary reactions to the immediate situation that are of no lasting significance.

This qualification highlights a distinctive feature of proposition C, which identifies the dyad as a context not merely of reciprocal interaction but of *reciprocal development*. It is from this point of view that the dyad, especially as it evolves into a primary relationship, constitutes a "developmental system"; it becomes a vehicle with a momentum of its own that stimulates and sustains developmental processes for its passengers as long as they remain interconnected in a two-person bond. Other, high-order interpersonal systems also exhibit this dynamic property but have additional features that introduce further complexity into the developmental equation.

Before turning to a consideration of these higher-order subsystems, I would like to emphasize the significance of the Western Reserve studies at a broader ecological level. Taken as a whole, this series of experiments on the effects of early, extended mother-infant contact provides excellent examples of several defining properties of an ecological research model, by both demonstration and default. On the positive side, the work constitutes a clear instance of ecologically valid experimentation focused directly on developmental processes. Moreover, it presents a fine illustration of how experimental intervention can bring to light critical features of an ecological process hardly likely to be identified through straightforward naturalistic observation in the unaltered existing setting. Last, but hardly least, the work provides an actually executed example of a transforming experiment. The investigators have deliberately and dramatically altered the established routine in American hospitals, clearly a macrosystem phenomenon. And they have done so "in ways that challenge the prevailing forms of social organization, belief systems, and lifestyles" (definition 11), in this instance at the level of the society as a whole. It is ironic that, at the same time, this series of studies exemplifies a striking ecological omission, a failure to take into account the actual system operating in the given environment.

This dramatic lacuna in an otherwise impressive body of research gives rise to the next proposition.

PROPOSITION D
An analysis of the microsystem must take into account the full interpersonal system operating in a given setting. This system will typically include all the participants present (not excluding the investigator) and involve reciprocal relations between them.

Once this proposition is formulated, it immediately suggests that perhaps the infant was not the only forgotten participant in the Western Reserve experiments: what about nurses, visitors, not to mention attending physicians, two of whom were apparently the principal investigators of the project? When these parties were present, as they surely were, did they act in similar fashion toward the mothers and infants in the experimental as compared with the control groups? Or did the lengthier and more intensive interaction between mother and newborn under the condition of extended exposure invite more approving comments to the mother about the infant and her way of handling him? Did the striking departure from the usual hospital routine lead the mothers to ask questions, and if so, how did the staff members respond?

As for family members or other visitors, it is noteworthy that, both in the United States and Guatemala, the experiments were conducted in hospitals serving primarily nonwhite populations from poor socioeconomic backgrounds. In the American controlled experiment (Kennell et al., 1974), the only one for which such background data are provided, of the fourteen mothers in each group all but one was black, two-thirds were unmarried, and all the children were first-born. Is it conceivable that the mothers' subordinate social status, cultural background, and (for most of them) their special position as a single mother of a first child, predisposed them to forming the strong kind of attachment they exhibited to their offspring? To put the same question in operational terms, would similar results have been obtained with a sample of white, middle class, two-parent families having their second or third child? The original investigators have acknowledged the importance of this issue.

The group of lower-class staff mothers has both advantages and disadvantages for a study of maternal attachment. The mothers had not been to childbirth classes, so they did not know what to expect in the hospital. They had done little reading, so they were rather "pure" for the purposes of this study. Almost all were black and their incomes and circumstances were similar in both groups. One difficulty with studies of maternal behaviour is that when people in the community begin to hear about it, their behaviour changes. Educated mothers may then behave in a special way because of what they have heard or read. (Kennell, Trause, and Klaus, 1975, p. 96)

In confining themselves to a two-person model, the Western Reserve investigators reflect yet another influence of the traditional laboratory paradigm. As previously noted, the classical psychological experiment allows for only two participants: E and S. Even in those researches that take into account the activities of more than two persons, the behavior of each is usually analyzed separately and interpreted as an independent effect. An example is provided by research on father-infant interaction.[1] Much of this work treats the behavior of the father, and any reaction it may evoke in the child, in exclusively class-theoretical terms (Lewin, 1935) as attributable entirely to the father, without regard to the possibility that both the father's action and the child's responses may be influenced by the mother—her presence, absence, and the possible effect of her behavior on the interaction of the father with the child. I refer to this kind of indirect influence as a *second-order effect*. To state the issue in propositional form

PROPOSITION E
In a research setting containing more than two persons, the analytic model must take into account the indirect influence of third parties on the interaction between members of a dyad. This phenomenon is called a *second-order effect*.

This proposition represents an extension and further specification of proposition D as applied to a system involving more than two persons, referred to henceforth as an $N + 2$ *system*. Three recent studies of parent-child interaction that, explicitly or implicitly, employed a three-person model illustrate the application of the principle. Parke (1978) and his coworkers observed both parents with their newborns in a hospital setting to determine what effect each parent had on the other's interactions with the infant. In each case

the presence of the spouse significantly altered the behavior of the other parent, specifically, both father and mother expressed more positive affect (smiling) toward their infant and showed a higher level of exploration when the other parent was also present ... These results indicate that parent-infant interaction patterns are modified by the presence of another adult; in turn, the implication is that we have assumed prematurely that parent-infant interaction can be understood by our sole focus on the parent-infant dyad alone. (Pp. 86–87)

Support for Parke's conclusion comes from a study by Pederson (1976), in which the second-order effect is somewhat more remote but equally consequential. This investigator examined the influence of the husband-wife relationship (assessed through interview) on mother-infant interaction in a feeding context (as observed in the home). His results are summarized as follows: "The husband-wife relationship was linked to the mother-infant unit. When the father was supportive of the mother ... she was more effective in feeding the baby ... High tension and conflict in the marriage was associated with more inept feeding on the part of the mother" (p. 6). Pederson also found that the developmental status of the infant, as measured on the Brazelton scale, was inversely related to the degree of tension and conflict in the marriage. Consistent with the principle of reciprocity, he notes that causality could occur in either direction.

Pederson's results indicate that the second-order effect can have inhibitory as well as facilitative impact. Indeed, Lamb interprets the results of three pioneering experiments explicitly designed to investigate second-order effects (Lamb, 1976b, 1977, 1978) as demonstrating that, beginning with the second year of life, the presence of the second parent reduces rather than increases parent-child interaction. His data do indeed show higher levels of interaction for a two-person parent-child system, but the interpretation is complicated by two problems of ecological validity. First, all the experiments were carried out in the laboratory. As I shall document below, a number of comparative studies (including one by Lamb) have shown that both parents and children behave rather differently in laboratory than in home settings. Second and more critical, the design employed in all three experiments involved a confounding of size of system with differing instructions given to the adult subjects about how they should behave toward each other as compared with the infant. Although they were asked to respond to the child's initiatives, the adults were enjoined from initiating interaction with the child but told to "chat to one another normally" (Lamb, 1977, p. 640). This directive meant that when an adult and a child were alone in the room, there was nothing to distract the former from reacting to the child's behavior. Once two adults were present, however, they were supposed to talk to each other. Thus their attention was focused on each other and drawn away from the infant. Under these circumstances, it is hardly surprising that adult-child interaction was lower in the three- than the two- person situation.

Lamb's interpretation of the observed difference as a second-order effect due solely to the presence of a second adult provides another example of failure to take cognizance of the actual interpersonal system operative in the setting (proposition D). It also illustrates the danger of artificially restricting the habitual behavior of research subjects, as is frequently done in laboratory experiments. While the results may be statistically reliable, they can also be experimental artifacts and hence ecologically invalid.

Again, this criticism does not mean that laboratory studies are necessarily suspect. When employed in proper ecological perspective, they often constitute the scientific strategy of choice. For example, if the laboratory is viewed as what it almost invariably is for a young child—namely, a "strange situation" (Ainsworth and Bell, 1970)—it reveals clearly the role of the parent as a source of security for the child and, in terms of a three-person model, as a catalyst for the child's interaction with the environment, including other, unfamiliar persons. Thus in all the strange-situation experiments, the mother's presence in the laboratory reduces the child's anxiety and resistance to the "stranger." The effect is even more pronounced in the home. For example, Lamb (1975, 1976c, 1977) finds that infants in the company of their parents look and smile at the stranger more often than at their mothers.

$N + 2$ systems and second-order effects of course occur in other settings. An instructive example from the school classroom is provided by Seaver (1973) who ingeniously exploited an "experiment of nature" to investigate the controversial phenomenon of induced teacher expectancies first reported by Rosenthal and Jacobsen (1968) and referred to by them as "Pygmalion in the classroom." Seaver's research was motivated by some reservations regarding the ecological validity of methods previously employed for the study of this phenomenon. In his words, "Most previous attempts to demonstrate the teacher expectancy effect have used experimental manipulations of teacher expectancies that were artificial and surely unusual in the experience of the teacher. Quite possibly these manipulations were also implausible to the teacher and induced psychological states other than the desired expectancies" (p. 334).

To achieve ecological validity, Seaver examined differences in the academic achievement of elementary school pupils with older siblings who had had the same teacher and performed either exceptionally well or exceptionally poorly. Children taught by teachers who had not instructed the older siblings served as controls. In contrast to earlier studies, which had produced inconsistent, weak, or questionable effects, the results of Seaver's natural experiment gave substantial support to the teacher-expectancy hypothesis. As Seaver himself acknowledged, however, it was not clear who was the mediator of the observed effect. Were the teacher's expectations changed because of her prior experience with the older sibling, or did the younger sibling evoke a different response from the teacher because of the younger child's expectations created by the older sibling or by the parents (based on their previous acquaintance with the teacher), or both? The remaining ambiguity in interpretation testifies to the importance of identifying and analyzing existing interpersonal systems and higher-order effects as stipulated in proposition E.

The involvement of one or both parents as intermediaries in a process already involving two siblings and a teacher would escalate the system from a triad to a quartet or quintet, or, more generally, an $N + 3$ system. To my knowledge, no empirical studies using such a model have been carried out, despite the fact that the so-called typical American family consisting of two parents and two children constitutes a readily available example.[2]

The family is at once the richest and most underused source of natural experiments on the developmental impact of $N + 2$ systems and second-order effects. In homes and families, one does not even have to introduce contrived variations in system size, for nature provides them on a daily basis. Parents and siblings—as well as relatives, neighbors, and friends—frequently come and go, providing ready-made experiments of nature with built-in ecological validity and a before-after design in which each subject can serve as her own control. The comings and goings are of two kinds. There are the temporary and recurring arrivals and departures, as adults and children go in and out of the room, friends and neighbors drop by, or—on a more predictable basis—family members leave for and return from work, school, and recreation, relatives come for a weekend or a week, or a parent gets a vacation from work. Then there are more lasting changes: a second child is born, grandma moves in to help with the children, mother goes off to work when the children are old enough (and "old enough" now comes sooner every year), grandmother dies and the family gets a regular babysitter, there is a separation or divorce and father leaves, after a few years the mother remarries, and so on.

Both temporary and lasting changes in system size can produce second-order effects. One can observe whether and how mother-child interaction changes when the father enters or leaves the room or how the total pattern of family activity is restricted when a second child arrives, mother takes a job, or father moves out. Given the frequency of such events, particularly in certain segments of contemporary society, one would expect that these experiments of nature would not have escaped scientific attention. But in an effort to find examples in the research literature, I have been able to discover only two studies employing the suggested strategy. The first is also one of the few investigations documenting the effect of an ecological transition *within* the family—the role change involved when a woman becomes the mother of a second child. The work was done over thirty years ago by a prescient leader in the field (Baldwin, 1947) and involved observations of maternal behavior toward the first child before, during, and after the mother's pregnancy with another child. Baldwin summarizes his results as follows: "All of these changes are linear in form. They suggest that the addition of another child in the family tends to reduce the warmth and contact between the parent and other children and to result in a more restrictive but less effective home" (p. 38).

Unfortunately, in keeping with a traditional research model Baldwin's research, like the Western Reserve studies, focused exclusively on one member of the dyad; data are provided only on the behavior of the parent and do not include that of the child. The rich scientific benefits to be gained by adopting a two-sided perspective are illustrated in the work of Hetherington and her associates.

Even though nearly half the children being born today will spend some time in a one-parent family, mostly as the result of separation or divorce (Glick, 1978), it would still require an extremely large sample to provide enough cases for a statistically adequate longitudinal study of the changes taking place in a family as it shifts from a three- to a two-person structure. Much can be learned, however, by observing the course of family life once the divorce has occurred, particularly if concurrent data are obtained for a matched sample of intact families. This was the strategy employed by Hetherington and her colleagues (Hetherington, Cox, and Cox, 1976, 1978) in a follow-up study of forty-eight recently divorced middle class parents in cases where custody had been granted to the mother. Divorced parents were identified and contacted through court records and lawyers. A comparison group of two-parent families was selected from a similar socioeconomic background and on the basis of having a child of the same sex, age, and birth order in the same nursery school as a child from a divorced family. In addition, an attempt was made to match parents with regard to age, education, and length of marriage. Only first- and second-born children were used in the study. The fact that both groups of families came from middle class backgrounds allowed the investigators to avoid a frequent source of confounding in studies of single-parent families, almost half of which (44 percent) have incomes below the poverty line (U.S. Bureau of the Census, 1977). The research procedures employed involved a wide variety of methods including parent interviews, observations of the parents and child interacting in the laboratory and home and of children's behavior in the nursery school, as well as checklists and ratings of child behavior provided by both parents and teachers. Measures were administered at two months, one year, and two years following the divorce.

In keeping with proposition C, developmental changes were found not only in the children but in the parents as well. Initially it was the fathers who were the hardest hit by the experience of separation. Feeling anxious, insecure, and inadequate, they engaged in a desperate search for a new identity in a variety of activities. But within a year the crisis had abated, primarily because they had established a new heterosexual relationship. The problems experienced by the mothers and the children had a longer course and were not so readily resolved. The following composite picture emerges from the rich and diversified data reported in the study.

Placed in the unaccustomed position of the family head, the mother often finds it necessary, because of her reduced financial situation, to look for work or a more remunerative job than her present one. At the same time, she must care for the house and children, not to mention create a new personal life for herself. The result is a vicious circle. The children, in the absence of a father, demand more attention, but the mother has other tasks that must be attended to. In response the children become more demanding. The data reveal that, in comparison with youngsters from intact families, the children of divorce are less likely to respond to the mother's requests. Nor does it make it any easier for her that similar requests are complied with when made by the divorced father. Even when the child is responsive to her, the divorced mother is less apt to acknowledge or reward the action. In the words of the authors,

Divorced parents made fewer maturity demands, communicated less well, tended to be less affectionate, and showed marked inconsistency in discipline and control of their children in comparison to married parents. Poor parenting was most apparent when divorced parents, particularly divorced mothers, interacted with their sons. Divorced parents communicated less, were less consistent, and used more negative sanctions with sons than with daughters. (1978, p. 163)

In keeping with the principle of reciprocity, the same pattern is mirrored in the behavior of the children toward their parents.

After reviewing the . . . findings one might be prone to state that disruptions in children's behavior following divorce are attributable to emotional disturbance in the divorced parents and poor parenting especially by mothers of boys. However, before we point a condemning finger at these parents, especially the divorced mothers who face the day to day problems of childrearing, let us look at the children . . . children of divorced parents exhibited more negative behavior than do children of intact families . . . These behaviors were most marked in boys and had largely disappeared in girls two years after divorce. Such behaviors were also significantly declining in the boys. Children exhibited more negative behavior with their mothers than with their fathers; this was especially true with sons of divorced parents.

The divorced mother was harassed by her children, especially her sons. In comparison with fathers and with mothers in intact families, the children of the divorced mother did not obey, affiliate, or attend to her in the first year after divorce. They nagged and whined, made more dependency demands, and were more likely to ignore her. (Pp. 169–170)

The disruptive effects of separation on parents, children, and their relations with each other reached their peak one year after the divorce and declined through the second year although the divorced mothers never gained as much control as their married counterparts. But given our criterion of developmental validity, the critical question is that of long-range effects. Is there any evidence that separation and divorce leave their mark on the behavior of the child in other settings and at other times? In a recent review of their own and other research on the development of children in mother-headed families, Hetherington and her colleagues concluded that "children living in mother-headed single-parent homes appear to be at higher risk for disruption in cognitive, emotional, and social development than are children in nuclear families" (Hetherington, Cox, and Cox, 1977, p. 31). The studies reviewed involved samples ranging in age from the early preschool years through adolescence and adulthood and varying widely in socioeconomic background.

Corroborative evidence from a broader and more systematic statistical base appears in findings from the National Survey of Children (Zill, 1978). Using a stratified probability sample of households in the United States containing at least one child in the age range of seven to eleven years, the investigators interviewed the eligible child and the parent who would be most capable of providing information about that child. Data on children of divorce exhibited a consistent pattern that prevailed after statistical control for socioeconomic status as measured by parental education and income. The general findings were summarized as follows:

Divorce significantly increases a child's risk of developing emotional and behavior problems. Children whose parents have been divorced by the time the child is of grammar school age are twice as likely to need or have gotten psychiatric help as children in intact families. Such children are more likely to have had a seriously disturbing experience, either due to the divorce itself, or to other life circumstances preceding or following the divorce. The minority of children who exhibit aggressive and antisocial behavior at home, in school, or at play, is larger among children of divorce than among children of intact families. Children of divorce are also more likely to feel neglected and rejected by their parents. (P. 53)

It is important to examine in greater detail the particular ways in which developmental disturbance is manifested among children from single-parent families. Relevant information is provided in the review by Hetherington and her colleagues (1977). In the social-emotional area, children from such families were likely to experience difficulties in sex role identification, show lack of self-control, and exhibit antisocial behavior. For boys, disruptions in sex role typing (as manifested by greater dependency, reduced levels of aggressiveness, and lower preference for masculine activities) tended to occur if separation from the father took place before the age of five. Differences were apparent from the preschool years onward, with some evidence of enduring effects through adolescence and young adulthood. For girls, differences did not emerge until adolescence and were concentrated in the area of heterosexual relationships. Women from homes in which the father had been absent had difficulty in establishing satisfactory relations with men. In general both men and women who had grown up in a single-parent family were more likely to experience marital instability than their counterparts from intact families, particularly if the single parent was female and the separation had been caused by divorce rather than death.

A similar pattern was found with respect to problems of self-control and antisocial behavior, with the additional feature that difficulties were considerably more pronounced if the children were male. It is the boy from a divorced home who is more likely to be impulsive, unable to delay immediate gratifications, inconsiderate, aggressive, or delinquent. Hetherington and her colleagues (1976, 1978) see this syndrome as a product of the especially antagonistic mother-son relationship observed in their divorced families.

The same line of interpretation is offered in explanation of the consistently poorer intellectual and academic performance of children, adolescents, and adults—especially males—brought up in homes broken by separation and divorce. Hetherington and her associates view the cognitive impairment as a product of disrupted socialization processes in the parent-child dyad. Drawing on the findings of their own study, they point out:

It was found that in divorced families there was a marked breakdown of appropriate and consistent parental control over children, fewer demands for mature independent behavior, and less communication, explanation and reasoning with children. These poor parenting practices were associated with high distractibility, impulsivity, short attention spans and lack of persistence on tasks by the children, which in turn were associated with drops in scores on performance and quantitative tasks and on certain types of problem solving tasks. Problem solving and academic success requires the ability to concentrate and persist. This ability to focus and sustain attention seems more critical in tasks that involve reasoning such as mathematical problem solving than on such things as vocabulary. Hence, the frequently reported quantitative-verbal discrepancy found in children in mother headed families . . . What is being proposed is that poor parental control leads to high distractibility and lack of persistence in children which causes poor problem solving performance. It would seem that the quality as well as the quantity of maternal interaction in single parent families should be considered. (1977, p. 13)

This conclusion is of course nicely in accord with the hypotheses I have offered regarding the conditions most conducive to human development. More precisely, it is the breakdown of these conditions that characterizes the mother-child dyad in divorced families, particularly during the first year after separation. Consistent with hypotheses 4 through 7, one sees the especially powerful disruptive impact on development of a mutually antagonistic primary dyad as the level of reciprocity diminishes, the intensity of negative interpersonal feelings increases, and the balance of power, instead of shifting gradually toward the child, becomes in the words of the mothers themselves a "declared war," a "struggle for survival," or "like getting bitten to death by ducks" (1976, p. 425).

It is regrettable that Hetherington and her colleagues do not provide an equally full account and analysis of the mother-child relation in two-parent families. The report contains no verbal descriptions or concrete examples for this group. One can only infer from the text and tables that mother-child dyads embedded in a three-person family system were characterized by a more effective socialization pattern. There was better communication between parent and child, and the mothers engaged in more explanation and reasoning, made more frequent demands for mature, independent behavior, showed greater consistency in discipline, and were more affectionate with their children. The youngsters themselves correspondingly exhibited more self-control, less antisocial behavior, a clearer sexual identity, more consideration for others, a greater capacity to defer gratification, and higher levels of intellectual and academic performance.

If these inferences are correct, they indicate second-order effects of impressive scope and consequence. It would appear that the presence of an adult with whom the mother has a positive relationship enables her to function more effectively in interactions with her child. Conversely, mutual antagonism in the husband-wife dyad, culminating in separation, disrupts the functioning of the mother-child dyad and impairs its capacity to serve as a context of effective socialization.

The impact of a third party on the functioning of an embedded dyad can be generalized in the form of a hypothesis that defines a key process and distinctive property of $N + 2$ systems.

HYPOTHESIS 8
The capacity of a dyad to function effectively as a context of development depends on the existence and nature of other dyadic relationships with third parties. The developmental potential of the original dyad is enhanced to the extent that each of these external dyads involves mutually positive feelings and the third parties are supportive of the developmental activities carried on in the original dyad. Conversely, the developmental potential of the dyad is impaired to the extent that each of the external dyads involves mutual antagonism or the third parties discourage or interfere with the developmental activities carried on in the original dyad.

The investigation of this hypothesis clearly requires the application of what is minimally a three-person model. Although Hetherington and her colleagues did not use such a model for analyzing the group in their sample to whom the triadic paradigm is most easily applied—two-parent families—, in their discussion of interpersonal relations in divorced families, they provide some elegant examples of how one dyad in a three-person system can be affected by the other two. Moreover, these examples are nicely in accord with directional processes stipulated in hypothesis 8.

The illustrations appear in the context of an examination by the investigators of exceptions to the general finding that there is disturbed psychological functioning among children from divorced families. The absence of such disturbance in the behavior of the child was associated with certain positive features in the mother-child dyad. For example, "children of divorced mothers who were available, who maintained firm but sensitive discipline, and encouraged independent mature behavior showed no cognitive deficits" (1977, p. 13). The investigators then raised the question of what factors might account for the capacity of these mothers to function effectively in dealing with their children. The most critical influence in this regard turned out to be the behavior of the divorced father and the relationship between the divorced parents:

Effectiveness in dealing with the child is related to support in child rearing from the spouse and agreement with the spouse in disciplining the child . . . When there was agreement in child rearing, a positive attitude toward the spouse, low conflict between the divorced parents, and when the father was emotionally mature . . . frequency of father's contact with the child was associated with more positive mother-child interactions and with more positive adjustment of the child. When there was disagreement and inconsistency in attitudes toward the child, and conflict and ill will between the divorced parents or when the father was poorly adjusted, frequent visitation was associated with poor mother-child functioning and disruptions in the children's behavior. (1976, pp. 425–426)

A similar influence on the effectiveness of the mother-child dyad was exerted by other third parties, but none of these was as potent as the primary relationship involving the father.

Other support systems such as that of grandparents, brothers and sisters, close friends, especially other divorced friends or male friends with whom there was an intimate relationship, or a competent housekeeper also were related to the mother's effectiveness in interacting with the child in divorced but not in intact families. However, none of these support systems were as salient as a continued, positive, mutually supportive relationship of the divorced couple and continued involvement of the father with the child. (P. 426)

This line of analysis leads the authors to a provocative conclusion. Having reviewed the research evidence on the problems experienced by divorced families, the processes involved, and their disruptive effects on the children growing up in these families, Hetherington and her associates address the crucial issue of causal factors, and arrive at what is essentially an ecological interpretation.

These developmental disruptions do not seem to be attributable mainly to father absence but to stresses and a lack of support systems that result in changed family functioning for the single mother and her children

... An increasing number of children are going to grow up in single-parent mother-headed families.

It is critical to develop social policies and intervention procedures that will reduce stresses and develop new support systems for single-parent families in order to offer these families [a] more constructive and fulfilling life style. (1977, pp. 31–32)

As the work of Hetherington and her coworkers demonstrates, looking beyond the mother-child dyad and applying an $N + 2$ model to the analysis of the family as a system inevitably directs the investigator's attention beyond relations in the immediate setting containing the child, or what we have called the microsystem, to influences emanating from successively more remote levels of the external environment—in our terms, the meso-, exo-, and macrosystems. Thus the capacity of the mother-child dyad to perform its developmental functions is seen to depend on the behavior not only of other members of the household but also of persons from the outside world. Some of these persons (such as a day care worker) interact with the child in other settings (mesosystem); others (such as a friend at work) may associate with the mother and never have contact with the child (exosystem); finally, as the investigators emphasize in their conclusion, the existence and nature of such external stresses and supports are in significant measure determined by the prevailing institutions and belief systems of the larger society (macrosystem). To effect any substantial change in the lives and thereby in the presently impaired psychological development of children from divorced families, it will be necessary to alter these existing institutional and ideological patterns.

I wish here to call attention to yet another signal aspect of Hetherington and her colleagues' outstanding investigation. It has to do neither with theory, substance, nor method but rather with an equally important requirement for effective scientific work—the initiative and resourcefulness of the investigators. This two-year longitudinal study of ninety-six divorced families and matched controls was conducted without any grant support while the principal investigator was carrying a full teaching load and editing a major research journal. Much of the planning was done in a graduate seminar and all the interviews and observations were conducted by student volunteers (E. M. Hetherington, personal communication). This is not to imply that substantial funds are not required for studying development-in-context, but it does demonstrate that original research on the ecology of human development can be carried out by workers who do not have massive financial resources and a paid staff to assist them. Two of the other studies on $N + 2$ systems discussed above were accomplished by young, individual investigators with only modest financial support (Lamb and Seaver).

In regard to the analysis of $N + 2$ structures within the microsystem, one issue remains to be considered, namely, the particular ways in which a third party can enhance or impair the capacity of a dyad to perform its developmental functions. We have already noted that a mother can serve as a source of security for an infant in relating to a stranger (Ainsworth and Bell, 1970) and as a reinforcer (and possibly a model) for the father in interacting with his newborn child (Parke, 1978). Conversely, the father's positive relation to the mother, especially in her child-rearing role, increases her effectiveness in the care and feeding of the infant (Pederson, 1976), enhances the quality of mother-child interaction (Hetherington, Cox, and Cox, 1976, 1977, 1978) and thereby fosters the child's psychological development (Hetherington, Cox, and Cox; Pederson). Similar positive effects are achieved by encouragement from relatives, neighbors, and friends (Hetherington, Cox, and Cox). Although the systematic evidence is still lacking, it appears likely that such persons can function constructively in a number of ways: serving as confidantes, aides, substitutes, or scapegoats, providing needed information, advice, or material resources, reinforcing initiatives, facilitating the formation of new social relationships, strengthening the power of a second person as a behavior model for the first (as when a mother praises her son when he acts like his father), or, as demonstrated in Seaver's research, creating expectations for how others should behave toward the child.

On the negative side, third parties can become sources of distraction (Lamb, 1976b, 1977, 1978), be perceived as rivals (Baldwin, 1947), or, as so graphically documented in the studies of Hetherington and her colleagues, impair the quality of primary relationships with the child through their own involvement in dyads with the other parent that are characterized by mutual hostility and frustration. A finding from the National Survey on Children (Zill, 1978) is significant in this regard. On indexes of psychological disturbance, there was one group of children that consistently obtained scores almost as high as those found for children of divorce. They were children of parents from intact families who on a three-point scale of marital happiness described their marriages as least happy. These families constituted 3 percent of all two-parent households. Their children were second only to those of divorced couples in the percentage reported by parents as experiencing psychological problems requiring professional help. Both parents and teachers also described these youngsters as among the more aggressive.

Even though the National Survey data are only cross-sectional and not longitudinal, the fact that children from divorced families typically showed an equal if not slightly higher level of psychological disturbance suggests that the legal separation of the parents did not bring about an improved situation for the child. This sobering result points to what is perhaps the most destructive effect of third parties on the course of human development—the damage produced by their absence. Such absence means the unavailability of someone to function in the constructive roles I have described, as in the case of a teen-age mother with a newborn having no one to whom she can turn for advice, assistance, encouragement, or mere companionship (Furstenberg, 1976).

The issue of the number of persons available as third parties to a given dyad calls attention to another distinctive property of an $N + 2$ system. Whereas the formation of a dyad, as I have defined it, requires that both participants be present in the same place at the same time, patterns of interaction in an $N + 2$ structure can be sequential. Many of the second-order effects described above are operative even though all the parties involved are not interacting simultaneously. The ex-husband, relative, or friend who offers support to the divorced mother in her child rearing role may do so when the child is not actually present. Such a sequential interaction system constitutes what I shall call a *social network*.

Since a minimum of three persons is required for a sequential interaction to take place, social networks are peculiar to $N + 2$ systems. A sequential interpersonal structure in which every member at some point interacts with every other member constitutes a *closed* social network.[3] A structure in which some theoretically possible dyads do not in fact occur is called an *incomplete* social network. Social networks can occur within a single setting, for example, in an office where certain employees are never present at the same time and have to communicate by leaving messages or through third parties. The most common and extensive social networks, however, are those that extend across settings and hence constitute elements of a meso- or exosystem. For this reason, I defer discussion of the properties of social networks, and their significance for development, to later chapters.

Important differences exist, documented by Hetherington and others (Felner et al., 1975; Hetherington, 1972; Hetherington, Cox, and Cox, 1977; Santrock, 1975; Tuckman and Regan, 1966), be-tween the development of children growing up in families in which the mother was widowed and in which she was divorced. The degree of disturbance, whether in the cognitive, emotional, or social realm, was consistently greater for the latter group than for the former. In assessing the factors contributing to this developmental difference, Hetherington and her associates (1977) point not only to the often continuing acrimony between divorced parents and the mother's anger at being abandoned but also to "the greater social stigma associated with divorce" and to the fact that "widows seem to have more extended support systems . . . than are available to divorcees" (p. 28). Consistent with this finding, results from the National Survey of Children (Zill, 1978) indicate a striking differ-ence in assessments of mental health for divorced as compared with widowed mothers. Whereas feelings of tension and depression are often reported by the former group, "widowed mothers, who are not much better off in terms of either education or income, are surprisingly free from psychological distress" (p. 24).

It would appear that, at least in American society, a single mother left with the care of a young child is treated differently depending on whether the marriage was ended by death or by divorce. As a result, women finding themselves in these positions are subject to different sets of pressures and react accordingly. This phenomenon shows the operation of another critical element of the microsystem —*social role*, which interests us as it functions to stimulate, main-tain, and, on occasion, dramatically redirect the course of human development.

5.

Roles as Contexts
of Human Development

The interpretations offered in the preceding chapter of the differences found in the behavior of children and mothers from single- as compared with two-parent families were based on the unstated assumption that the observed effects were attributable to the different social positions (married versus divorced) occupied by the mothers and the role expectations associated with these positions. An alternative interpretation exists, one that views the divorced status of the mother as an outcome rather than a cause, the product of personality maladjustment that existed prior to the marriage and led to the intrafamilial conflicts that culminated in legal separation. According to this point of view, behavior problems in the mother-child relationship of the kind documented by Hetherington and her colleagues would have been present before the divorce and hence could not be explained as reflecting the differential impact of a two- as opposed to a three-person system. Moreover, the finding that children whose mothers were widowed rather than divorced showed less psychological disturbance, rather than being viewed as consistent with a role hypothesis could be regarded as being in accord with a personality-oriented interpretation: mothers who divorced were, and had been, maladjusted; those whose husbands died, were not—they were simply victims of fate, unselected in other respects.

The last assertion would be difficult to maintain given the fact that widowed mothers tend to be older and more well-to-do than those who are divorced. Yet recognition of this difference would not demolish the personality-oriented explanation. To do so would require random assignment of future parents to one or another marital status, a prospect that constitutes a perfect example of an experimental manipulation that can never be carried out in modern

83

civilized societies, for ethical as well as practical reasons. At least one hopes so.

Ethical considerations notwithstanding, it has been possible to make role assignments at random in other kinds of real-life situations, and the results constitute dramatic evidence that placing people in different roles, even in the same setting, can radically influence the kinds of activities and relations in which they engage and thereby presumably alter the course of their development. I say "presumably" because, in keeping with a conventional research model, virtually all the experiments conducted to date are confined to a single setting and a limited period of administration; hence there is no evidence as to the continuity of the experimentally induced changes over place and time, so that the criterion of developmental validity remains unfulfilled. It would be quite unwise to assume, however, that experiences of the nature and intensity occurring in these experiments would not, if continued over a longer interval, have some lasting effect that carries over beyond the research situation.

One other caveat is in order. While not going so far as to determine on a chance basis whether and when a person should marry or divorce, some experiments do create a situation in which persons selected at random are subjected to profoundly disturbing emotional and social experiences of an intensity not anticipated by the scientists who conducted the experiments. As a result, serious questions have been raised about the justifiability of such experiments from the viewpoint of the ethics of science. I not only share some reservations on this score but take the position that the failure to recognize the potential of psychological damage from experiments of this kind derives in part from the limitations of the conventional research model, which fails to look beyond consequences to the individual subject while she is in the research setting. As a result, even the most conscientious investigator can overlook the possibility of effects on the same person in other settings, or on her relation to other people in her life (children, spouse, parents, friends, and so on) and thus on these significant others themselves, even though the research subject may remain unaffected and unaware of this effect. It is therefore conceivable and to be hoped for that, had the original researchers been exposed to and employed an ecological model in the design of their experiments, the potential dangers would have been recognized and avoided.

It is necessary here to clarify what is meant by the concept of role as employed in the present theoretical framework. An ecological approach requires some modification of the generally accepted definition of role as "the behavior expected of the occupant of a given position or status" (Sarbin, 1968, p. 546). Whereas this definition does imply a phenomenological frame of reference, it fails to take into account the element of reciprocity central to the systems orientation being developed here and indeed included in the classical formulations of the construct of role by G. H. Mead (1934) and Cottrell (1942). These original conceptions encompass not only expectations about how a person in a given social position is to act toward others but also how others are to act toward that person (thus when a teacher explains, the pupil is expected to pay attention). In terms of microsystem elements, these can be characterized as expectations about reciprocal activities and relations. Accordingly, our definition of role incorporates all these features.

DEFINITION 14
A *role* is a set of activities and relations expected of a person occupying a particular position in society, and of others in relation to that person.

Roles are usually identified by the labels used to designate various social positions in a culture. These are typically differentiated by age, sex, kinship relation, occupation, or social status, although other parameters (such as ethnicity and religion) may also come into play. Operationally, a person's social position and hence her role label can be defined as a reply to the question, "Who is that person?" from the perspective of someone acquainted with both the person and the social context in which the person is located.

Associated with every position in society are *role expectations* about how the holder of the position is to act and how others are to act toward her. These expectations pertain not only to the *content of activities* but also to the *relations* between the two parties, in terms of the dyadic parameters previously outlined: degree of reciprocity, balance of power, and affective relation. The contrasting roles of parent and teacher are examples. Both are expected to provide guidance to the young, who in turn are expected to accept such guidance in a relation characterized by high levels of reciprocity, mutual affection, and a balance of power in favor of the adult. But with parents the degree of reciprocity and mutual affection is presumed to be higher, and parental authority is thought to extend over a broader segment of the child's life than the teacher's, at least in modern Western societies.

It is clear that the concept of role involves an integration of the elements of activity and relation in terms of societal expectations. Since these expectations are defined at the level of the subculture or culture as a whole, the role, which functions as an element of the microsystem, actually has its roots in the higher-order macrosystem and its associated ideology and institutional structures.

It is the embeddedness of roles in this larger context that gives them their special power to influence—and even to compel—how a person behaves in a given situation, the activities she engages in, and the relations that become established between that person and others present in the setting.

The special power of roles is elegantly demonstrated in an experiment by Zimbardo and his colleagues (Haney, Banks, and Zimbardo, 1973), entitled "Interpersonal dynamics in a simulated prison." A concise description of the research design and process is provided in a review by Banuazizi and Movahedi (1975).

The study . . . was conducted in the summer of 1971 in a mock prison constructed in the basement of the psychology building at Stanford University. The subjects were selected from a pool of 75 respondents to a newspaper advertisement asking for paid volunteers to participate in a "psychological study of prison life." The 24 subjects who were chosen were male college students, largely from middle-class backgrounds, who were judged by the experimenters to be the "most stable (physically and mentally), most mature, and least involved in anti-social behaviors [Haney et al., 1973, p. 73]." On a random basis, half of the subjects were assigned to the role of guard and half were assigned to the role of prisoner.

Prior to the experiment, the subjects were asked to sign a form with the following stipulations: (a) All subjects would agree to play either the prisoner or the guard role for a maximum of two weeks: (b) those assigned to the prisoner role should expect to be under surveillance, to be harassed, and to have some of their basic rights curtailed during their imprisonment, but not to be physically abused; and (c) in return, the subjects would be guaranteed a minimally adequate diet, clothing, housing, medical care, and financial remuneration at the rate of $15 per day for the duration of the experiment.

One day before the start of the experiment, the guards were invited to an orientation meeting. They were informed that the goal of the study was to "simulate a prison environment within the limits imposed by pragmatic and ethical considerations," and that their task was "to maintain the reasonable degree of order within the prison necessary for its effective functioning." The prisoner subjects were telephoned and asked to be available at their homes on a given Sunday at which time the study would begin. Subsequently, with the cooperation of the Palo Alto City Police Department, the subjects were apprehended in a "surprise [?] mass arrest." After going through an elaborate arrest and booking procedure, the subjects were blindfolded and driven to the mock prison.

Although the authors did not have any specific hypotheses to test, the general purpose of the study was to explore the interpersonal dynamics of a prison environment through a functional simulation of a prison in which no prior dispositional differences existed between prisoners and guards, and each group played its respective role for a maximum of two weeks.

The outcome of the study was quite dramatic and not entirely expected by the authors. In less than two days after the initiation of the experiment, violence and rebellion broke out. The prisoners ripped off their clothing and their identification numbers and barricaded themselves inside the cells while shouting and cursing at the guards. The guards, in turn, began to harass, humiliate, and intimidate the prisoners. They used sophisticated psychological techniques to break the solidarity among the inmates and to create a sense of distrust among them. In less than 36 hours, one of the prisoners showed severe symptoms of emotional disturbance, disorganized thinking, uncontrollable crying and screaming, and was released. (Later, there was a rumor that he had been faking and had won his release under false pretenses.) Soon the prisoners asked that a grievance committee be established and church services provided. On the third day, a rumor developed about a mass escape plot, which prompted the guards and the superintendent (Professor Zimbardo), who was operating in the background, to take various preventive measures. The guards, in the meantime, increased their harassment, intimidation, and brutality toward the prisoners. On the fourth day, two prisoners showed symptoms of severe emotional disturbance and were released, while a third prisoner developed a psychosomatic rash all over his body. He was also released. On the fifth day, the prisoners showed symptoms of individual and group disintegration. They had become mostly passive and docile, suffering from an acute loss of contact with reality. The guards, on the other hand, had kept up their harassment, some behaving sadistically and "delighting in what could be called the ultimate aphrodisiac of power [Zimbardo et al., 1972]." (P. 153)

The main findings of the study are summarized by the original authors as follows:

Continuous, direct observation of behavioural interactions was supplemented by video-taped recording, questionnaires, self-report scales and interviews. All these data sources converge on the conclusion that this simulated prison developed into a psychologically compelling prison environment. As such, it elicited unexpectedly intense, realistic and often pathological reactions from many of the participants. The prisoners experienced a loss of personal identity and the arbitrary control of their behaviour which resulted in a syndrome of passivity, dependency, depression and helplessness. In contrast, the guards (with rare exceptions) experienced a marked gain in social power, status and group identification which made role-playing rewarding.

The most dramatic of the coping behaviour utilized by half of the prisoners in adapting to this stressful situation was the development of acute emotional disturbance—severe enough to warrant their early release. At least a third of the guards were judged to have become far more aggressive and dehumanizing toward the prisoners than would ordinarily be predicted in a simulation study. Only a very few of the observed reactions to this experience of imprisonment could be attributed to personality trait differences which existed before the subjects began to play their assigned roles. (Haney, Banks, and Zimbardo, 1973, p. 69)

Banuazizi and Movahedi have challenged the validity of the foregoing conclusions on the ground that, from a strict methodological perspective, the behavior of the experimental subjects is correctly interpreted not as the product of what Zimbardo and his colleagues had called "a psychologically compelling prison environment" but as a conforming response to expectations communicated by the investigators in the experimental situation. In the critic's own words:

To account for the behavioral outcomes of the experiment, we offer the following alternative explanation: (a) The subjects entered the experiment carrying strong social stereotypes of how guards and prisoners act and relate to one another in a real prison; (b) in the experimental context itself, there were numerous cues pointing to the experimental hypothesis, the experimenters' expectations, and possibly, the experimenters' ideological commitment; and thus (c) complying with the actual or perceived demands in the experimental situation, and acting on the basis of their own role-related expectancies, the subjects produced data highly in accord with the experimental hypothesis. (P. 156)

In support of their argument, Banuazizi and Movahedi submit results of a questionnaire administered to a sample of college students. After reading a description of the experimental situation used by the Stanford group, the respondents were asked to speculate about the nature of the major hypothesis being tested and the probable behavior of subjects assigned as "prisoners" and "guards." The results revealed that "the overwhelming majority of the respondents (81%) were able to articulate quite accurately the intent of the experiment, that is, its *general* hypothesis" (p. 157). In addition, although there was some variability in predictions about the anticipated reaction of prisoners, there was high consensus about the guards; 90 percent of the respondents predicted that they would be authoritarian, oppressive, and hostile toward the prisoners.

In short, the issue being raised is one of ecological validity. In the terms of our definition of this concept, the charge is that the environment experienced by the subjects did *not* have the properties it is supposed or assumed to have by the investigator.

Readers of the critique by Banuazizi and Movahedi were quick to respond. With respect to the thesis that the subjects' behavior was the product of social stereotypes, a number of workers in the field (DeJong, 1975; Doyle, 1975a; Thayer and Sarni, 1975), emphasized that the same argument applies with equal force to persons who in real life find themselves for the first time in the position of being a guard or a prisoner; they too begin to act in accord with the cultural stereotypes existing for these roles in a given society.

Moreover, whereas there was consensus among the students in Banuazizi and Movahedi's sample about the probable behavior of guards, there was no such agreement about the likely reaction of prisoners; the respondents had split evenly in predicting "rebellious," "passive," and "fluctuating" patterns of response. Also, as one reviewer pointed out (DeJong, 1975), Banuazizi and Movahedi themselves had stated that the original authors "did not have any specific hypotheses to test" (p. 153); indeed, Zimbardo had indicated in an article on the ethics of the Stanford experiment (1973) that had the experimenters anticipated the extreme reactions that occurred, the study would never have been run.

De Jong calls attention to the changes in the behavior of the research subjects over time. As the experiment continued, the guards intensified their harassment and aggressive behavior, whereas the prisoners exhibited marked shifts in emotional state. On the second day, they broke out in violence and rebellion, but once the guards used force to quell the uprising (by employing fire extinguishers as weapons and transforming prisoners' rights into "privileges"), there followed a period of profound passivity, depression, and self-depreciation.[1] While social stereotypes, fed by accounts of prison violence in the mass media, might well have played a part in these sequential reactions, there was clearly a dynamic process at work.

Given the cogency of this rebuttal, it seems untenable to dismiss the results of the Stanford experiment on the ground that they are experimental artifacts. At the same time, Banuazizi and Movahedi have presented convincing evidence for the operation of social stereotypes. How does the interpretation offered by Zimbardo and his colleagues stand up in the light of the established existence of such stereotypes?

In my view, contrary to Banuazizi and Movahedi's, the new evidence only provides more empirical support for the original authors' theory. The influence of social stereotypes fits nicely as one of several factors in their implicit conceptual model, which, as Banuazizi and Movahedi correctly point out, is nowhere fully explicated. The underlying assumptions can be identified, however, from the investigators' statement of the rationale for the prison experiment.

According to its originators, the Stanford study was undertaken to challenge "a prevalent non-conscious ideology: that the state of the social institution of prison is due to the 'nature' of the people who administer it, or the 'nature' of the people who populate it, or both." The authors refer to this widely shared cultural orientation as the *dispositional hypothesis,* the view that "a major contributing cause to despicable conditions, violence, brutality, dehumanization, and degradation existing within any prison can be traced to some innate or acquired characteristic of the correctional and inmate population" (Haney, Banks, and Zimbardo, p. 70).

In opposition to this individualistic orientation, Zimbardo and his associates offer what is, in effect, an ecological explanation of the behavior of both the prisoners and their guards. They propose that the reactions of the experimental subjects represented primarily not the manifestation of enduring personality characteristics but rather patterns of response specific to particular roles and institutions in contemporary American society. In the investigators' view, the only way to demonstrate the validity of this alternative hypothesis was to carry out a controlled experiment.

[A] critical evaluation of the dispositional hypothesis cannot be made directly through observation in existing prison settings, since such naturalistic observation necessarily confounds the acute effects of the environment with the chronic characteristics of the inmate and guard populations. To separate the effects of the prison environment *per se* from those attributable to *a priori* dispositions of its inhabitants requires a research strategy in which a "new" prison is constructed comparable in its fundamental social-psychological milieu to existing prison systems, but entirely populated by individuals who are undifferentiated in all essential dimensions from the rest of society. (P. 71)

In my view, the results of the experiment, even if conservatively interpreted, provide unequivocal evidence for the operation of the "prison" environment in producing the observed effects. The question then becomes what aspects of that environment were at work, and how they functioned. The argument and evidence adduced by Banuazizi and Movahedi make a strong case for the operation of role stereotypes existing in the society at large, but the cogent rebuttal of the countercritics shows that this is not the whole story. In particular, the critique fails to account for the sequential changes in the behavior and emotional state of the subjects over time, especially those seen in the "prisoners," for whose role no consistent expectations apparently existed. These temporal developments do have a place, however, in the interpretative schema offered by the investigators themselves. In their view, the critical factor in the experiment was the creation and implied social sanction of a power relation between prisoner and guard, a relation that then pursued an inexorable momentum of its own.

The dynamics of this relation are indicated in the following excerpts from the investigators' report.

The conferring of differential power on the status of "guard" and "prisoner" constituted, in effect, the institutional validation of these roles ... Being a guard carried with it social status within the prison, a group identity (when wearing the uniform), and above all, the freedom to exercise an unprecedented degree of control over the lives of other human beings. This control was invariably expressed in terms of sanctions, punishment, demands and with the threat of manifest physical power ... The use of power was self-aggrandizing and self-perpetuating ... Guard aggression showed a daily escalation even after most prisoners had ceased resisting and prisoner deterioration had become visibly obvious to them ... The prisoner participation in the social reality which the guards had structured for them lent increasing validity to it and, as the prisoners became resigned to their treatment over time, many acted in ways to justify their fate at the hands of the guards, adopting attitudes and behavior which helped to sanction their victimization ... The typical prisoner syndrome was one of passivity, dependence, depression, helplessness and self-deprecation. (Pp. 89–94)

Three elements can be distinguished in this line of interpretation. The first is role legitimation, accomplished by setting the role in the context of institutions firmly established in the society. One of these institutions is clearly identified—the existing prison system; the other is conspicuously used but not explicitly acknowledged —the university. The conduct of the experiment at Stanford under the aegis of the psychology department clearly lent the approval and authority of scholarship and science to the "prison," its administrators, and its guards. The second structural element involved investing the guards with legitimate authority. As a result they exercised their power over the prisoners in ever-escalating fashion. Third, the ultimate response of the prisoners was to accept the submissive dehumanized role defined for them by the behavior of their guards.

Several initial hypotheses regarding the impact of role allocation on behavior can now be formulated.

HYPOTHESIS 9

The placement of a person in a role tends to evoke perceptions, activities, and patterns of interpersonal relation consistent with expectations associated with that role as they pertain to the behavior both of the person occupying the role and of others with respect to that person.

HYPOTHESIS 10

The tendency to evoke perceptions, activities, and patterns of interpersonal relation consistent with role expectations is enhanced when the role is well established in the institutional structure of the society and there exists a broad consensus in the culture or subculture about these expectations as they pertain to the behavior both of the person occupying the role and of others with respect to that person.

HYPOTHESIS 11

The greater the degree of power socially sanctioned for a given role, the greater the tendency for the role occupant to exercise and exploit the power and for those in a subordinate position to respond by increased submission, dependency, and lack of initiative.

A more felicitous, if less precise, phrasing of this principle is of course found in Lord Acton's classic aphorism: "Power tends to corrupt and absolute power corrupts absolutely" (1948, p. 364).

The interpretation of the results of the Stanford experiment in terms of the dynamics of power and submission, as well as the results themselves, inevitably call to mind Milgram's (1963, 1964, 1965a, 1965b, 1974) much-discussed research on "obedience to authority," referred to by some, including the late Gordon Allport, as the "Eichmann experiment" (Milgram, 1974, p. 178). The following summary of the design and principal findings is taken from a review by Aronson and Carlsmith (1968):

In these experiments, Milgram asked subjects to give a series of electric shocks to a person, ostensibly as part of a learning experiment. Unknown to the subject, no shocks were actually dispensed. After each "incorrect" trial the subject was asked to increase the intensity of the shocks by pressing one of a continuous series of levers labeled from "Slight Shock" at one end to "Danger: Severe Shock" near the other end. The majority of the subjects continued to increase the shock level to the maximum in spite of the fact that the "recipient" (actually a confederate) who was closeted in the next room indicated that he was in severe pain, pounded on the door, and finally fell silent. Since the confederate's silence constituted an incorrect response on the "learning" task, the subjects were asked to increase the intensity. The majority obeyed. Milgram provided a vivid description of the effects this procedure had on the typical subject who complied with the experimenter's requests. His behavior included sweating, stuttering, profuse trembling, uncontrollable nervous laughter, and, in general, an extreme loss of composure. (Pp. 22–23)

As Zimbardo and associates acknowledge, their findings and conclusions echo Milgram's earlier work in demonstrating the proposition that "evil acts are not necessarily the deeds of evil men, but may be attributable to the operation of powerful social forces" (Haney, Banks, and Zimbardo, 1973, p. 90). But the Stanford group point out correctly that their own research goes beyond Milgram's in clarifying a number of important substantive and theoretical issues.

Most significant was a critical difference in experimental design. In the Milgram study, the behavior of the authority figure was completely predetermined and limited to a series of escalating commands. As a result, the responses of the subject were correspondingly limited essentially to compliance or resistance. In the Stanford experiment, "instructions about how to behave in the roles of a guard or prisoner were not explicitly defined" (Milgram, 1974, p. 91) so that members of either group "were essentially free to engage in any form of interaction" (p. 80). Yet their emergent encounters turned out to be overwhelmingly hostile and sharply polarized along the dimension of authority versus submission. This fact, together with the evolutionary changes that occurred over the six-day period of the experiment, constitutes much more powerful evidence of the capacity of role structures per se to motivate and *shape* the course of behavior than was provided in Milgram's research.

At the same time, Milgram's investigation yielded results that not only support and clarify conclusions reached by the Stanford experiment but call attention to powerful social influences receiving little attention in the reports by Zimbardo and his colleagues. As to the former, a modest modification within the setting produced a substantial effect. Observing that the scientific experimenter's role possesses a "status component," Milgram arranged that the experimenter, who was to give commands for administering shocks to the victim, be called away for a phone call just before the experiment was to begin. An accomplice, who appeared to be another subject awaiting his turn, then volunteered and actually carried out exactly the same procedure used in the basic experiment. As a result, "there was a sharp drop in compliance . . . only a third as many subjects followed the common man as follow the experimenter" (p. 97). Milgram concludes, "It is not what subjects do but for whom they are doing it that counts" (p. 104).

Consistent with the results of the Stanford experiment, this finding pins down the authority status of a role as a critical element in producing conformity. An additional aspect of Milgram's research identifies yet another factor affecting the capacity of roles to elicit behavior. Working from the perspective of Asch's studies of social conformity in perception (1956), Milgram explored the effect on the subject's response of the presence and behavior in the experimental situation of other persons who served as confederates either of the experimenter or of the subject himself. He found that obedience to the experimenter increased if a confederate encouraged the subject to administer a higher shock (1964) and decreased if confederates opposed the shock (1965a).

This additional finding leads to a further hypothesis regarding environmental conditions influencing the tendency of roles to evoke behavior in accord with social expectations. The underlying principle is already somewhat familiar, since it represents an extension to the level of roles of hypothesis 8, (which deals with the influence of third parties in an $N + 2$ system).

HYPOTHESIS 12
The tendency to evoke behavior in accord with expectations for a given role is a function of the existence of other roles in the setting that invite or inhibit behavior associated with the given role.

The confederates in Milgram's staged learning experiment were of course deliberately introduced into the setting by the investigator and were acting in accordance with a predetermined script. But in group situations outside the laboratory, as in Zimbardo's simulated prison, such confederates could emerge spontaneously on one side or the other. One can therefore ask what interpersonal structures, if any, developed in the mock prison? The hostility between the prisoners and the guards was marked, but what about friendships or antagonisms within these two groups? Finally, moving beyond the dyad to the level of $N + 2$ systems, we can ask: to what extent did third-party relations mediate the polarized interaction between prisoners and guards?

Zimbardo and his colleagues provide surprisingly little information relevant to these matters. Although sociometric measures were apparently administered to all subjects (Haney, Banks, and Zimbardo, 1973, p. 77), no results are reported. Data on the prisoners' interactions are confined to the content of private conversations, monitored by audio recording in the yard and in the prison cells. The investigators were surprised to discover the delimited nature of these exchanges:

When the private conversations with the prisoners were monitored, we learned that almost all (a full 90%) of what they talked about was directly related to immediate prison conditions; that is, food, privileges, punishment, guard harassment, etc. Only in one-tenth of the time did their conversations deal with their life outside the prison. Consequently, although they had lived together under such intense conditions, the prisoners knew surprisingly little about each other's past history or future plans . . . The guards, too, rarely exchanged personal information during their relaxation breaks. They either talked about "problem prisoners," or other prison topics, or did not talk at all. (P. 92)

Although this account suggests that few intimate or intense relationships developed within either group, it is apparent that some members did talk with each other, since some kinds of interpersonal structures did in fact emerge. As early as the second day the prisoners organized a rebellion and, even after it was forcibly put down, set up an elected grievance committee. The fact that $N + 2$ systems were actually operative in the prison setting is reflected in the investigator's discovery, from postexperimental interviews, "that when individual guards were alone with solitary prisoners and out of range of recording equipment, as on the way to or in the toilet, harassment was often greater than it was in the 'Yard'" (p. 92).[2]

The occurrence of such events focuses attention on the desirability of systematically examining the extent to which the behavior of prisoners and guards varied as a function of their involvement in dyads, cliques, or quasi-formal roles and structures (such as membership in the grievance committee). The absence of such an analysis in the Stanford study highlights a paradoxical aspect of this important transforming experiment. Even though it provides eloquent evidence in support of an ecological model, it exhibits a number of limitations characteristic of traditional research paradigms. The underlying conceptual framework is essentially limited to a two-person model (guard and prisoner) with no allowance for the existence or emergence of $N + 2$ structures. Even more striking and consequential is the fact that the investigation is completely confined to the level of the microsystem.

Data obtained on the effects of the experimental treatments were restricted almost entirely to responses exhibited by the subjects while in the simulated prison. Although some kind of assessment was carried out with each subject some months after completion of the experiment, the nature of the inquiry is not specified, and the only reference to results is a single and somewhat cryptic statement: "Follow-ups on each subject over the year following termination of the study revealed the negative effects of participation had been temporary, while the personal gain to the subjects endured" (p. 88). No systematic information is available about the prior or subsequent experience of the subjects at home, in school, on the job, or in other settings, nor about their relations with family members, friends, fellow students, coworkers, and so on.

From an ecological perspective, such information is significant on several counts. If it is true—as proposed in hypothesis 8—that within a microsystem third parties can have a profound impact on interpersonal relations, then it seems plausible that such an influence could extend across as well as within settings. In the Stanford experiment, the existence and nature of interpersonal linkages with persons outside the mock prison could well have mediated the character and degree of reaction manifested by the subjects. Did the students who had strong ties with their parents, with a wife or sweetheart, or with a peer group respond any differently to the experimental situation from those lacking or experiencing frustration in such close interpersonal relationships? Such questions remain unanswered, since the only background data examined, or at least thus far reported, by the Stanford group have been scale scores on personality inventories, which, consistent with the investigators' "anti-dispositional" orientation, accounted for "an extremely small part of the variation in reaction to this mock prison experience" (p. 81).

Analogous considerations apply to the Milgram experiment. For example, does the subject's readiness to obey or resist the commands of the experimenter vary as a function of his present (or past) relation with authority figures in the home or on the job? Questions of this kind remain unanswered and, for that matter, unasked in both studies. Later I shall examine some empirical evidence from other studies suggesting that, had the Stanford and Yale investigators pursued these issues, they would have obtained results of theoretical and social importance.

Furthermore, increased attention to ecological questions would make possible a sophisticated consideration of the ethics involved in such research. From a traditional laboratory perspective, it seems quite sufficient, in considering the possible harmful effects of a particular experiment, to focus on the well-being of the experimental subject. An ecological orientation, however, places at the center of attention not just the individual but the interpersonal systems in which he participates both within and across settings. Of particular importance in this regard are primary dyads typically involving parents, spouse, children, and friends. All the subjects in both the Zimbardo and Milgram experiments had relatives or companions. In the Stanford study, some of them actually visited the mock prison and presumably heard from the prisoners at first hand about their painful experiences, or even directly observed the effects of such experiences in the appearance and behavior of persons to whom they were closely attached. And surely, when both experiments were over, the subjects must have talked about what they had been through to their families and other close associates. Given the intensity of the experience for the subjects in both investigations, one cannot rule out the possibility that, even without their being aware, they were affected in their subsequent interactions with others, particularly in the domains of dominance, submission, and response to authority. These are, of course, dimensions that pervade relationships in the family, the job, and, indeed, every aspect of life. In sum, it is not inconceivable that experiments of the kind conducted by Zimbardo and Milgram affect the well-being of others besides those who actually serve as the research subjects and whose consent to involvement in the scientific enterprise was neither sought nor obtained.

Such a possibility considerably extends the realm of responsibility incumbent on investigators of human behavior and development. Maintaining an ecological perspective may broaden the researcher's view of the possible human impact of scientific activities and result in greater care and restraint in the design of experiments having the potential to affect adversely the well-being not only of the research subject but of the other persons in his life.

The issue posed by the Stanford and Yale experiments that is most central to the ecology of human development is the following: whether the intense emotional reactions observed in both experiments—reactions that often reached a pathological level—have any enduring impact on the subjects' behavior after their return home, especially in their relations with family members, friends, coworkers, superiors or subordinates. The real-life equivalent of this query highlights its significance for public policy: does the incarceration of human beings in real prisons and concentration camps have any lasting impact on their behavior after release in their relations with family members, friends, coworkers, superiors, or subordinates?

Unfortunately, because of the ecologically constrained conceptual model employed in both experiments, the question remains unanswered. In fact to my knowledge, no systematic research on these important developmental issues exists. It would be difficult, but obviously not impossible, to obtain relevant information from real prisoners, but for the same reasons that Zimbardo and his colleagues undertook their experiment, the results would be confounded by the preselection of the prisoner populations. Such confounding could still have been avoided, however, and relevant information obtained much more easily, by follow-up interviews with the research subjects in the Stanford and Yale experiments, and perhaps with their close associates as well. Most efficient would have been a before-and-after design permitting assessment of changes in interpersonal relations at a point shortly following release and at successive intervals thereafter. The use of such a research strategy, however, presupposes a theoretical model that extends beyond a single locale to include transitions and interactions between settings.

Although the long-range effects of the Stanford and Yale experiments remain unknown, these investigations provide eloquent evidence of the power of social roles to induce behavioral change. The behavior thus produced is hardly reassuring from a societal perspective and also begs a question from a theoretical viewpoint: can roles be created that evoke constructive orientations rather than extremes of authoritarianism, submission, and psychological disorganization?

An earlier, equally ingenious and elegant, research classic deals with this issue. In the early 1950s, Sherif and his colleagues at the

University of Oklahoma conducted a unique experiment (1956, 1961). In the words of Elton B. McNeil,

War was declared at Robbers Cave, Oklahoma in the summer of 1954 . . . Of course, if you have seen one war you have seen them all, but this was an interesting war, as wars go, because only the observers knew what the fighting was about. How, then, did this war differ from any other war? This one was caused, conducted, and concluded by behavioral scientists. After years of religious, political, and economic wars, this was, perhaps, the first scientific war. It wasn't the kind of war that an adventurer could join just for the thrill of it. To be eligible, ideally, you had to be an eleven-year-old middle-class, American, Protestant, well-adjusted boy who was willing to go to an experimental camp. (1962, p. 77)

Sherif set out to demonstrate that within the space of a few weeks, he could bring about two sharply constrasting patterns of behavior in this sample of normal boys. First, he would transform them into hostile, destructive, antisocial gangs; then, within a few days, change them again, this time to become cooperative, constructive workers and friends concerned about and even ready to make sacrifices for each other and for the community as a whole.

The success of the effort can be gauged by the following excerpts describing the behavior of the boys after each stage had been reached. After the first experimental treatment was introduced,

good feeling soon evaporated. The members of each group began to call their rivals "stinkers," "sneaks," and "cheaters." They refused to have anything more to do with individuals in the opposing group. The boys . . . turned against buddies whom they had chosen as "best friends" when they first arrived at the camp. A large proportion of the boys in each group gave negative ratings to all the boys in the other. The rival groups made threatening posters and planned raids, collecting secret hoards of green apples for ammunition. In the Robbers Cave camp, the Eagles, after a defeat in a tournament game, burned a banner left behind by the Rattlers; the next morning the Rattlers seized the Eagles' flag when they arrived on the athletic field. From that time on name-calling, scuffles and raids were the rule of the day . . . In the dining-hall line they shoved each other aside, and the group that lost the contest for the head of the line shouted "Ladies first!" at the winner. They threw paper, food and vile names at each other at the tables. An Eagle bumped by a Rattler was admonished by his fellow Eagles to brush "the dirt" off his clothes. (Sherif, 1956, pp. 57–58)

But after the second experimental treatment,

the members of the two groups began to feel more friendly to each other. For example, a Rattler whom the Eagles disliked for his sharp tongue and skill in defeating them became a "good egg." The boys stopped shoving in the meal line. They no longer called each other names, and sat together at the table. New friendships developed between individuals in the two groups.

In the end the groups were actively seeking opportunities to mingle, to entertain and "treat" each other. They decided to hold a joint campfire. They took turns presenting skits and songs. Members of both groups requested that they go home together on the same bus, rather than on the separate buses in which they had come. On the way the bus stopped for refreshments. One group still had five dollars which they had won as a prize in a contest. They decided to spend this sum on refreshments. On their own initiative they invited their former rivals to be their guests for malted milks. (P. 58)

How was each of these effects achieved? Treatment I has a familiar ring, at least in American society: "To produce friction between the groups of boys we arranged a tournament of games: baseball, touch football, a tug-of-war, a treasure hunt and so on. The tournament started in a spirit of good sportsmanship. But as it progressed good feeling soon evaporated" (p. 57).

But how does one turn hatred into harmony? Before undertaking this task, Sherif wanted to demonstrate that, contrary to the views of some students of human conflict, mere interaction—pleasant social contact between antagonists—would not reduce hostility: "We brought the hostile Rattlers and Eagles together for social events: going to movies, eating in the same dining room and so on. But far from reducing conflict, these situations only served as opportunities for the rival groups to berate and attack each other" (p. 57).

Conflict was finally dispelled by a series of stratagems. Sherif gives the following example: "Water came to our camp in pipes from a tank about a mile away. We arranged to interrupt it and then called the boys together to inform them of the crisis. Both groups promptly volunteered to search the water line for trouble. They worked together harmoniously, and before the end of the afternoon they had located and corrected the difficulty" (p. 58). On another occasion, just when everyone was hungry and the camp truck was about to go to town for food, it turned out that the engine wouldn't start, and the boys had to pull together to get the vehicle going.

According to Sherif the critical element for achieving harmony in human relations is joint activity in behalf of a *superordinate goal.* "Hostility gives way when groups pull together to achieve overriding goals which are real and compelling for all concerned" (p. 58).

The transformation of goals in Sherif's experiment was achieved through a different process from that involved in the Stanford research. The key strategy in the Stanford experiment was to place subjects in familiar roles already existing in the larger society and representing extremes along a continuum of power. The constructive roles into which the campers were cast in the Robbers Cave experiment had no readily recognizable counterparts in the external world and were differentiated not in terms of authority status but rather by the content and aim of the activities to be pursued. But just as in the case of contrasting power positions, the differing aims and activities, once established, exhibited a momentum of their own and generated distinctive patterns of interpersonal structure which in this instance were characterized by harmonious as well as disruptive human relationships.

The strategy and its resultant dynamics can be generalized in the form of a hypothesis.

HYPOTHESIS 13
The placement of persons in social roles in which they are expected to act competitively or cooperatively tends to elicit and intensify activities and interpersonal relations that are compatible with the given expectations.

The Robbers Cave Experiment was not the first to demonstrate that arbitrary role assignments could produce socially constructive as well as deleterious effects. Analogous results were achieved in what can rightfully be called the prototype of all transforming experiments, of which both Sherif's and Zimbardo's studies are direct descendants—Lewin, Lippitt, and White's classic work at the University of Iowa on the creation and consequences of three contrasting leadership styles in children's groups (Lewin, Lippitt, and White, 1939; Lippitt, 1940; White and Lippitt, 1960). The design and findings have been summarized by Getzels (1969):

[The researchers] examined the effect of three leadership roles and the consequent group climates by observing the behavior of four "clubs" of five 10- or 11-year-old boys, each under three leadership conditions: "democratic," "autocratic" ("authoritarian"), and "laissez-faire." Groups were matched to control for individual differences and leaders were rotated to control for treatment variation. Records were kept of relevant behavior, including the interaction within the group, between the leader and individual boys, the expression of aggression, and productivity in club projects. One observation stood out above all others. The social climates resulting from the different leadership styles produced significant differences, which can be briefly summarized as follows: (1) aggressive behavior was either very high or very low under authoritarian conditions, extremely high under laissez-faire conditions, and intermediate under democratic conditions; (2) productive behavior was higher than or as high in authoritarian climates when the leader was present as in democratic climates but much lower when the leader was absent, moderately high and independent of the leader's presence or absence in the democratic climates, and lowest in the laissez-faire climates. (P. 505)

From an ecological perspective, the Iowa experiment was ahead of its successors not only in years but also in theoretical scope. Although no data were obtained regarding possible effects of the experimentally induced group climates on the subjects' subsequent behavior in other settings, the researchers did examine the influence of each boy's experience in a prior context on his reaction to the experimental conditions. They explored the extent to which parent-child relations in the home affected the boys' behavior in the three contrasting leadership styles. This phase of the inquiry was stimulated by the considerable individual variation exhibited by boys in response to a given type of leader, particularly when he took a democratic approach. Some youngsters made democracy difficult to achieve; others made it easier. The critical factor turned out to be a cluster of intercorrelated personality characteristics including honesty, modesty, perseverance, and nonaggression. The investigators referred to this set of qualities collectively by the term *conscience,* defined as "the psychological processes that lie back of the traits in this cluster and that account for the intercorrelations between them" (White and Lippitt, 1960, p. 200). Ratings of interviews conducted with the boys' mothers, as well as less systematic case studies, pointed to the conclusion that "the parents whose children become conscientious are likely to be those with the greatest warmth of affection and the greatest firmness or consistency—not severity—of discipline" (p. 221).

Of the experiments on role transformations we have examined thus far, this is the only one to seek and find a connection between events in the primary research setting and in some other area of the child's life. Nor is this the only case in which an early experiment recognized and explored important theoretical issues neglected in later work. The experimental designs employed by Lewin, Lippitt, and White in the thirties and by Sherif and his associates in the fifties (published 1961) made provisions for documenting a phenomenon of developmental significance implied but not explicitly demonstrated in the subsequent work both of Milgram and of Zimbardo and his colleagues.

In the Yale and Stanford studies, the contrasting roles of experimenter and subject, and guard and prisoner, were taken by different people. In the Lewin and the Sherif experiments, the *same* persons found themselves successively in different roles; they underwent a *role transition*. A shift of this kind represents one form of a more general phenomenon earlier defined as an ecological transition, occurring "whenever a person's position in the ecological environment is altered as the result of a change in role, setting, or both" (Definition 6).

In the Lewin and the Sherif experiments, the setting remained constant, but the subjects were successively placed in different roles, with corresponding alterations both in their own behavior and in their treatment by others. The developmental significance of such shifts in role becomes more readily apparent when they occur not in the context of a typically short-lived experimental situation but as changes of status in real-life settings; for example, the arrival of a younger sibling transforms the previously only child into an older brother or sister, a pupil is left back to repeat the same grade, a wife and mother becomes a single parent, an employee is promoted to supervisor. Given the research findings reviewed above, there is little question that such role transitions result in marked alteration of behavior. Although these findings arose from spontaneously occurring "experiments of nature," with each subject serving as his own control in a built-in before-after design, the demonstration that role transition has led to significant modifications of behavior in the same setting does not per se constitute evidence of developmental change. In keeping with the criterion of developmental validity, it is necessary to establish that the change carries over to other settings at other times, for example, that after a new sibling arrives in the home, the child begins to act differently in school, or that the husband's promotion at work alters his behavior as parent and spouse.

Similarly, it is the issue of developmental validity that underlies my emphasis on the importance of conducting follow-up studies for the Stanford, Yale, and Oklahoma experiments in order to determine whether the artifically induced, brief, but intensive experiences of being a prisoner, guard, heartless experimenter, cutthroat competitor, or cooperative fellow camper had any effect on the subsequent behavior of the subjects in their personal lives at home, at school, or on the job. In the absence of such information, none of these studies of role transformation meets the criterion of developmental validity. They could have been modified to do so by only a modest extension of the research design that would have provided minimal before-and-after data about behavior in at least one other setting. The expansion of scientific inquiry in this fashion would of course escalate the research model to the level of what I have called the mesosystem. Many role transitions in fact occur at this level, since they involve a change in setting as well as in social position; examples of these are entering day care, getting promoted to the next grade, graduating, changing jobs, and retiring.

The fact that in the Lewin and Sherif experiments the same children were able to play very different roles raises the question whether and how exposure to a variety of roles can affect the course of development. There appears to be no empirical evidence bearing on this issue. The basic theses of role theory as developed by G. H. Mead (1934), the Thomases (1927, 1928), Sullivan (1947), and Cottrell (1942) would seem to imply that such exposure should facilitate the process of psychological growth. All these theorists view personality development as the outcome of a process of progressive role differentiation involving two complementary phases. In the first instance, the child's psychological growth is faciliatated by his interaction with persons occupying a variety of roles—first within the home (mother, father, siblings, grandparents) and then beyond (peers, teachers, neighbors, and so on). At the same time, as a function of exposure to persons in different social positions, the child herself is constantly in new roles and develops a more complex identity as she learns to function as a daughter, sister, grandchild, cousin, friend, pupil, teammate, and so on. This general formulation gives rise to the following hypothesis.

HYPOTHESIS 14
Human development is facilitated through interaction with persons who occupy a variety of roles and through participation in an ever-broadening role repertoire.

If this hypothesis is valid, it gives cause for concern about the constricted range of roles currently found in the two primary settings of socialization in American society—the home and the classroom. At least since the late 1940s there has been an accelerating reduction in the number and size of extended families containing children (Bronfenbrenner, 1975) as well as a rocket rise in the proportion of single-parent families, with the latest census figures (U.S. Bureau of the Census, 1978) indicating that about one child in five is living with only one parent. With respect to classrooms, I have commented elsewhere (Bronfenbrenner, 1978b), on the paucity of adult models available to school children in the United States and other Western societies by contrast with a number of other cultures, notably the Soviet Union (1970a) and the People's Republic of China (Kessen, 1975). This increase in role exposure is best achieved not by increasing already overburdened school staff but by exposing pupils to adult roles existing in the larger society, both through bringing such persons into the school setting and through involving the children in activities in the outside world.

PART THREE

The Analysis of Settings

6.

The Laboratory
as an Ecological Context

The operation of the microsystem as a totality is reflected in a proposition that constitutes a basic tenet of the ecological approach.

PROPOSITION F
Different kinds of settings give rise to distinctive patterns of role, activity, and relation for persons who become participants in these settings.

This statement does not mean that there is no continuity in the behavior of an individual from one setting to the next but only that such continuity is accompanied by systematic differences. From the perspective of everyday experience, the proposition is so obvious as to appear trivial in its consequences. For example, it implies that a child will act somewhat differently at home and at school or that his father or mother do not behave in the same fashion on the job as within the family. Yet in developmental research, the same almost self-evident proposition often remains unrecognized, unresearched, and unheeded, thus creating problems in interpreting and generalizing findings. Although the overwhelming bulk of research on human development has been conducted in the laboratory setting, very little is known about this setting as a context for assessing behavior and development. What little is known should give us pause.

The most systematic comparisons of the laboratory with other settings focus on contrasting patterns of parent-infant interaction in the laboratory and the home. An example is Ross, Kagan, Zelazo, and Kotelchuck's (1975) comparative study of the same developmental phenomenon—separation protest—induced experimentally in two different settings. The usual paradigm employed for this purpose is the so-called strange situation (Ainsworth and Bell, 1970; Ainsworth, Bell, and Stayton, 1971; Ainsworth and Wittig, 1969; Bell, 1970, Blehar, 1974; Rheingold, 1969b; Rheingold and Ecker-man, 1970). In this procedure the child is brought to a laboratory room with his mother, and sometimes the father as well (Kotel-chuck et al., 1975; Lester et al., 1974; Spelke et al., 1973). After a brief period of becoming familiarized with the surroundings, in-cluding a strange adult who is also present, the parent goes away, and the infant is left alone with the stranger for a few minutes. Observations are made of the degree of distress and protest ex-hibited by the child after separation and upon the parent's return. Ross and her colleagues wondered what would happen if the strange situation was made at least partially familiar by moving it out of the laboratory into the home.

To answer this question, they carried out the strange situation experiment in both settings with two comparable groups of mother-infant pairs. The babies ranged in age from twelve to eighteen months and were all from middle class families. In the homes the experiment was usually conducted in the living room, or in the kitchen if the living room was not suitable. The authors summarize their findings as follows:

Although infants in both home and laboratory were maximally upset by being left alone with the stranger but minimally upset when left alone with either parent, there was significantly less distress in the home than in the unfamiliar laboratory ... Children tested in the laboratory cried almost three times as long as those tested at home during the two condi-tions when they were left alone with the stranger ... The greater dis-tress in the laboratory was also reflected in the significant ... change in duration of play for the conditions following adult departures ... Playing decreased twice as much in the two periods following the mother's de-parture in the laboratory than at home. (P. 256)[1]

In the following year Brookhart and Hock (1976) investigated the same phenomenon but used a different design. The strange sit-uation experiment was repeated with the same children in both settings but counterbalanced for order so that half the sample was initially observed in the laboratory and the other half in the home. Comparison of the infants' responses in the two settings revealed a pattern similar to that found by Ross and her associates. "As the episodes progressed, a greater increase in the intensity of contact-maintaining behavior occurred in the laboratory as compared to that in the home. In the laboratory, separation from mother and being left alone elicited more clinging and demands for prolonged physical contact than did similar events in the home" (p. 337). The authors conclude with a methodological caution:

The character of the settings within which behavior is observed clearly influences the infants' behavior with their mothers and with strangers. In designing studies of infant behavior, methodological attention should be given to the context in which behaviors are observed, and findings should be interpreted with systematic regard to contextual variables and their influence on observed behavior. Also, the limits of generalizability of findings across settings should be specified. (P. 339)

The attenuation of positive reactions by infants in a "scientific" environment is also illustrated in Lamb's comparative study (1975) of mother-father-infant interaction in home and laboratory settings. An eight-month-old baby's tendency in the home to display more behaviors (for instance, looking, smiling, reaching, vocalizing) toward the father disappeared in the laboratory. Instead the infants exhibited a greater desire for proximity to the mother. In a cautionary note, Lamb points out that "most ... studies that have reported preference for the mother over the father have relied on observations in strange laboratory situations" (p. 4).

Another finding by Lamb has significance in this regard. Sroufe (1970), Tulkin (1972), and others have contended that the university laboratory represents an unfamiliar and hence anxiety producing situation for lower class families. Consistent with this view, Lamb found that socioeconomic differences in father-infant interaction favoring middle class groups appeared in the laboratory situation, whereas they had not been present in the home. Only in the former setting did infants from lower class families look at their fathers less often and vocalize less frequently than their counterparts from middle class homes.

The social class difference found only in the laboratory involved the behavior of eight-month-old infants. There are two possible explanations for such a difference. The first is that a laboratory setting already begins to take on different meanings to children from contrasting social class backgrounds when they are eight months old. Perceptive as babies become during the first year of life, such social precociousness seems highly unlikely. The more plausible explanation is that the infants were reacting to the behavior of their parents, which varied as a function of their social class position. Unfortunately Lamb's study provides no data to support or refute this interpretation, since in keeping with the classical research model the only object of scientific attention was the experimental subject, in this instance the infant. Even if the observer's focus had been broadened to include the parent-child dyad, only a much-constricted view of the parental side of the interaction would have been obtained, since in line with the variable- as opposed to systems-oriented paradigm of the laboratory, the investigator sought to control presumed extraneous factors by instructing the parents not to initiate interaction.

A study by Schlieper (1975) cites data consistent with the conclusion that social class differences in parent-child interaction are more likely to appear in the laboratory than in the home. In an observational study of three- and four-year-olds and their mothers in their own homes, the investigator found "very few differences . . . between the low and middle SES mothers" (p. 470). Schlieper contrasts her findings with the marked social class differences reported by researchers observing families in the laboratory (Kogan and Wimberger, 1969; Walters, Connor, and Zunich, 1964; Zunich, 1961) and other university settings (Bee et al., 1969). In these latter studies, the low SES mothers consistently emerged as less stimulating and interactive than their middle class counterparts. Schlieper concludes:

Zunich and co-workers not only found many more differences between groups, but also found differences in the opposite direction from the present study. Their low SES mothers were less directing and less likely to play interactively, while in the present study they were more likely to show these behaviors. Zunich's lower-class group was much more likely to be "out of contact" with the children, while the present study showed no differences in this category. Zunich observed his subjects, who seem to have been very similar to those in the present study, at a university laboratory. (P. 470)

This interpretation of social class differences departs from the usual treatment of such data in the research literature by suggesting that the observed differences may be specific to the laboratory setting. At the same time, however, the interpretation follows an established pattern in describing the differences as an indication of less adaptive patterns of response on the part of lower class parents in the laboratory situation. The ecological perspective suggests another possibility, namely, that the appearance of social class differences in the laboratory also reflects a shift in the behavior of the middle class parents. Results consistent with this interpretation came from two comparative investigations with middle class samples.

In an observational study of twenty three-year-old children from the Merrill-Palmer Nursery School and their mothers, Schalock (1956) compared mother-child interaction in a home with that in a laboratory playroom and found significant differences in twenty-one out of thirty categories of maternal behavior. The nature of these differences is summarized as follows:

Directing the child by command was used often by the mother in the home, relatively infrequently in the playroom. Mother and child participated jointly in activities considerably more often in the home (78 episodes) than in the playroom (19 instances). While the mother showed 397 instances of Non-attention in the home, she failed to follow the child's behavior in the playroom in only 30 instances. The mother occasionally forbade, restricted, and criticized the child in the home, but only rarely in the playroom. (Moustakas, Sigel, & Schalock, 1956, p. 132)

Similar results are reported by Belsky (1976), who compared the behavior of twenty-four mothers and their one-year-old infants in the laboratory and in the home. Social class was controlled by including only families in which one parent had graduated from college or both had had some college experience. The observations revealed differences according to setting in the behaviors of the mothers but not of their children: "The general level of maternal functioning was greatly affected. Mothers attended to, spoke to, responded to, stimulated and praised their children more frequently in the laboratory than at home. In addition, they were more likely to ignore their children in the home and tended to prohibit them more and praise them less in this setting than in the laboratory" (p. 13).

Belsky interprets his own findings as indicating a tendency for mothers to exhibit more socially desirable behavior in the laboratory setting. Two other ecological influences, however, may also have been operative. First, as Belsky points out, the mother in the home may have been more occupied with and distracted by other activities, whereas in the laboratory she was free to devote all her attention to the child. This differential tendency was in all likelihood exacerbated by the fact that, in Belsky's experimental design, setting and experimental instructions were confounded. Whereas in the home mothers were told to "disregard the presence of the observer" and "go about their daily routines," in the laboratory they were invited "to pretend they were at home in the same room with their child with a half-hour of free time on their hands" (p. 5).[2] Nevertheless, the similarity of Belsky's findings to those of Schalock suggests that with comparable instructions the same class differences would have appeared, perhaps in a somewhat attenuated form.

Besides a possible tendency for mothers to behave more "properly" in the university setting, another explanation for the observed differences needs to be considered: the mother's greater involvement with her infant in the laboratory may have been prompted by motives reflecting her sensitivity to the potential effect on her infant of strange and unfamiliar surroundings. Thus the fact that the mothers "attended to, vocalized to, and responded to their infants more often in the laboratory" and exhibited more frequent expression of "positive affection" (p. 15) may reflect a desire to forestall their baby's distress rather than simply to show off their own child-rearing skills. Consistent with this interpretation is the fact that, despite the greater stimulation and responsiveness exhibited by the mothers in the laboratory, the infants showed no corresponding increase in vocalization or in other forms of response to the mother's initiative. On the contrary, nondistress vocalizations were significantly more frequent in the home than in the laboratory. Moreover, an examination of Belsky's data reveals that in the home the child vocalized significantly more often at the time of the second visit, whereas no corresponding gain appeared in the laboratory. This pattern persisted over time despite the fact that mothers tended to be somewhat more attentive to their infants during the second laboratory session than the first, although showing the opposite trend in the home.

Thus Belsky's results appear to be consistent with the interpretation that the mother's greater solicitousness in the laboratory situation reflects her awareness of and represents an adaptive response to the potentially disturbing impact of a strange situation upon a young child.[3] Her actions are quite appropriate when seen from her perspective. One is reminded of Harry Stack Sullivan's discerning comment about the appropriateness of the behavior of schizophrenic patients, given their perception of the situation: "We are all more simply human than otherwise" (1947, p. 7).

Given two explanations for the higher level of attention shown to the child in a laboratory as opposed to a home situation, is there a way to establish the independent operation of one or the other presumed influence? One possibility is to eliminate the presence of the observer as a salient figure in the setting, thus reducing any incentive for the mother to show off her maternal motivation and skills. This approach was used in an experiment by Graves and Glick (1978), who undertook to investigate the hypothesis that, in a setting in which the subjects know that they are being observed by a researcher, white middle class mothers in particular will seek to convey their image of a "good socializer" who stimulates the child, engages in a high level of verbal interaction, and provides instruction without appearing to demand.

The research subjects were six mother-child pairs, all from white middle class backgrounds. The children ranged in age from one and a half to two years. The experiment was conducted in two identical rooms at the university equipped as playrooms. Both contained a comparable set of toys, picture books, blocks, a tumbling mat, as well as a one-way mirror and videotape equipment. The experimental procedure was as follows:

> Prior to their arrival, mothers were told that the study involved videotaping of the mother-child dyad at play, and that the exact nature of the study and the variables being considered could not be disclosed until completion of the sessions. Each of the mothers agreed to these conditions.
> Upon arriving, the mothers and children were shown the two rooms where the taping was to take place. Each subject pair was videotaped for fifteen minutes in each of the two rooms consecutively. In the "observed" condition, subjects were aware that they were being taped—the videocamera, manned by an experimenter, was set up just outside the doorway of the room. In the "unobserved" condition, subjects were not aware that they were being videotaped. Throughout the unobserved session, the experimenter appeared to be in the process of setting up equipment with a colleague outside the door to the experimental room, while subjects actually were being videotaped through the one-way mirror in their room. (P. 43)

All mother-child pairs were exposed to both experimental conditions, counterbalanced for orders. The results revealed "striking differences" in the behavior of the mothers, with those in the "observed condition" being much more active.

First, the mothers approximately doubled the amount of speech they had produced when unaware of the observation. Their utterances were a bit shorter in general length and almost entirely directed toward the child. Here, the mothers appeared to be working avidly at teaching their children new words and skills, and at getting them to display those they had already acquired. The elicitation of the child's performance was virtually continuous. Sometimes the video camera and the experimenters themselves were the subjects of discussion and investigation by mother and child. Mothers also frequently repeated their prior utterances, giving the child a better chance to process the language and respond appropriately, and the child often was reinforced with a positive evaluative comment when he answered a question correctly or performed a manual task with facility. Mothers in general seemed very involved in the interaction with their children when they were aware that they were being taped. Naming or action "games," where a joint focus of attention was maintained between mother and child, occurred nearly 85 percent of the time. Mothers often initiated interactions with their children intentionally, and were almost always responsive to the child's own attempts at initiating interaction. (P. 45)

Like previous investigators of the effects of the research context on experimental performance, Graves and Glick view their results as having serious implications for the interpretation of scientific data.

These findings pose a problem for some of the work on mother-child interaction that attempts to characterize the nature of the interaction without consideration of the experimental social context which seems to affect behavior so radically. It must be taken into account that a psychological study where measurement is not unobtrusive, whether done in the home or in the laboratory, is a specialized context in which mothers try to put their "best foot forward" according to what they feel is appropriate and advantageous in this particular situation. Generalization from a limited set of observations about the overall developmental milieu for a child, or children in general, seems unwarranted . . . In cases when the experimenter shares a similar socioeconomic background, there may be a match between mother's and experimenters' expectations of what should constitute this role. If backgrounds differ, so may expectations and strategies for dealing with the world. Whichever the case may be, all analyses of mother-child interaction need to attend to contextual variables that might be affecting the nature of the interaction displayed. (P. 45)

It should be noted that the conditions of this experiment differed in an important respect from those obtaining in the previous studies we have examined, all of which involved the presence of a stranger in the laboratory setting. Hence Graves and Glick's clear-cut demonstration of the mother's effort to make a good impression on the observer does not rule out the possibility that greater maternal protectiveness is expressed toward the young child in the laboratory environment when the circumstances invite such a response.

The interpretation of the mother's behavior in the laboratory as an adaptive response to a definition of that situation consistent with her maternal role finds support from a comparative study by O'Rourke (1963). His experiment complements and clarifies the researches examined thus far in three respects. First, he focused on changes across settings in the behavior of fathers as well as mothers. Second, the sample consisted of families with older children (teen-agers). Third, and probably most consequential in terms of scientific yield, he approached the comparison of behavior in home and laboratory from a theoretical perspective.

Extrapolating from Bales's (1955) analysis of adaptive versus integrative functions in groups, O'Rourke hypothesized that "as the physical and/or social environmental conditions in which the group must function become less familiar, the adaptive problem of the group is heightened. For family groups this means that, as they move from their homes to an unfamiliar situation like the interaction laboratory, the problem of adaptation will come to have priority for solution over that of integration" (p. 425). Operationally, O'Rourke predicted that, in terms of Bales's interaction categories, the same families would in general exhibit more positive socioemotional behavior in the home than in the laboratory. He then carried the process of theoretical deduction several steps further to posit a highly complex pattern of differential reaction to the setting transition on the part of mothers versus fathers in relation to daughters versus sons. O'Rourke's starting point was the thesis of Parsons (1955) and Zelditch (1955) "that males are assigned and socialized to primarily instrumental-adaptive roles in the family, while expressive-integrative primacy is assigned to female members" (O'Rourke, p. 424). In the light of social norms associated with the two settings and two sex roles considered jointly, O'Rourke reasoned that a shift to the more instrumental challenge of the laboratory should evoke more positive responses from mothers, especially when in the presence of their daughters, but a decrease for fathers, especially when with their sons.

To test these hypotheses, he observed twenty-four three-person family groups in the home and in the laboratory. Half the groups included a male teen-age son, and the other half an adolescent daughter. In both settings, the families engaged in the completion of group projective tasks and the discussion of two decision-problems.

An analysis of the interaction data yielded support for all the proposed hypotheses: "As groups move from the familiar environment of the home to the unfamiliar laboratory situation, there is a general increase in instrumental and negative socio-emotional behaviors. These are the changes we would expect when the adaptive problem of a group is intensified" (p. 434). As predicted, the greatest rise in negative behavior from the home to the laboratory was shown by fathers of boys, whereas mothers of girls shifted in the opposite direction, expressing more positive feelings in the laboratory than in the home. It will be recalled that a similar differential pattern was exhibited by the mothers of preschool children in the studies by Schalock (1956) and Belsky (1976). No major interaction effects by sex of child were reported in these studies, perhaps because the children were still too young to evoke a parent's full differential response to their identity as a male or female person.

O'Rourke ends his report on a now-familiar cautionary note:

It is clear that researchers must become acutely aware of the differences in behavior which are elicited by the laboratory itself. On the basis of the results presented here, we must conclude that groups seen only in the laboratory will experience more disagreement among members, will be more active but less efficient at decision making, and will register less emotionality than might be the case if they were seen in their "natural" environments. Consequently, the laboratory situation works a definite distortion on the experimental outcomes. (P. 435)

Whereas O'Rourke's caveat is certainly justifiable, from an ecological viewpoint his broad statement warrants two qualifications. First, the substantive differences he describes may be specific to the contrast between the laboratory and the home in particular rather than "natural environments" in general. Second, from our theoretical perspective, the laboratory is a setting like any other—a place where people can readily engage in face-to-face interaction. Whether it "works a definite distortion on experimental outcomes" (O'Rourke, p. 435) depends on what kind of environment the laboratory situation is presumed to represent and to what other kinds of situations the experimental findings are generalized. In short, the ecological validity of a setting, be it a laboratory or a locale in real life, is never a forgone conclusion.

Nevertheless the laboratory does have its special vulnerabilities, particularly when results are taken as applicable to everyday life. The scientific and social risk is greatest when performance in an artificial and ephemeral situation is used as a basis for making judgements about individuals, institutions, or public policies. Blehar (1974) compared the reactions of forty two- to three-year-olds and their mothers in a "strange situation." The experiment was carried out in an unfamiliar room at the university. Half the children had been enrolled in full-time day care and half had been raised at home. When left with the stranger, the day care group exhibited greater distress and were more resistant and hostile both toward the stranger and toward the mother upon her return. Blehar interpreted her findings as indicating qualitative disturbance in the mother-child relationship in day care children and thus as having implications for policy and practice.

Given the evidence of differences in reaction to separation from the mother in home versus laboratory settings, such broad inferences seem premature. *Mother-child relationship* implies an enduring, generalized pattern of reciprocal feelings and acts. It remains to be demonstrated that the disturbance and antagonism shown by a child upon being left by his mother with a stranger in a strange setting provide any indication of the quality of the mother-child relationship in the home or other familiar locale. Especially if the results are to be used as a basis for determining public policy, research on the effects of day care on the mother-child relationship, or any other aspect of the child's life in the real world, is most appropriately conducted in the actual settings in which children live.

Blehar's data and design have been challenged in a replication of her experiment by Moskowitz, Schwartz, and Corsini (1977). Noting that in the original study no measures had been taken to insure that observers and coders were "blind" either to the research hypotheses or, more important, to the children's prior exposure to day care, these investigators controlled for such possible bias by videotaping the subjects' behavior for subsequent coding by persons unfamiliar with either the purposes or the design of the study. Under these more stringent conditions, the results not only failed to support but in at least one respect even contradicted Blehar's findings. Moskowitz and her colleagues found no differences in behavior toward the mother on the part of children with versus children without prior experience in day care. In terms of reaction to the experimental situation as a whole, the day care children showed less distress than their individually matched controls. Three other well-designed replications of the Blehar experiment (Brookhart and Hock, 1976; Doyle, 1975b; Portnoy and Simmons, 1978) found no significant differences as a function of mode of care.[4]

The Blehar experiment and its replications are very instructive. At the most general level, they indicate that as research extends beyond the sphere of basic science into the realms of practice and policy, the requirements for rigorous experimental design and theoretical analysis become even more critical. At a more substantive level, these experiments underscore the special significance of the laboratory as an ecological context.

The consistent pattern of differences in behavior exhibited in the laboratory as opposed to the home by both children and parents indicates something of what the laboratory experience means for these groups of subjects. For them, and perhaps for most people, the laboratory is indeed a strange situation that tends to evoke a person's characteristic response to the alien and the unfamiliar. This phenomenon pertains especially to young children but also to members of low income groups, ethnic minorities, and rural populations —perhaps to everyone except those who are or have been college students.

There are, of course, a good many of the latter. For them the laboratory usually means a place to make a good impression on the "scientist" or perhaps to outwit her at her own game. But as Milgram's experiment so soberingly demonstrates, even the sophisticated, finding themselves in a situation divorced from the rest of their lives, can unexpectedly depart from their habitual self-control into deviant extremes. As a result behavior evoked in the laboratory, especially when the researcher's aim is to reveal human frailties, is likely to exaggerate any maladaptive responses the same subject would have made in a real-life situation.

Piliavin, Rodin, and Piliavin (1969), noting that much of the research on giving help to a victim had been conducted in the laboratory, carried out a field experiment on "good Samaritan" behavior in the New York subway. On the seven and one-half-minute run between 59th and 125th streets on the West Side, the investigators "staged standard collapses" of a victim who appeared either ill (carrying a cane) or drunk (carrying a bottle). The original purpose of the study, to test the effect of a model in activating helpful behavior, was frustrated by the frequency and rapidity of spontaneous help offered by the passengers. In almost 80 percent of the trials, someone came to the rescue before the model could act. In the words of the investigators, "The frequency of help received by the victims was impressive, at least as compared to earlier laboratory results . . . On the basis of past research, relatively long latencies of spontaneous helping were expected; thus, it was assumed that models would have time to help, and their effects could be assessed" (p. 292). In this real-life situation of clear need, people turned out to be quite helpful. Why did they not act similarly in the laboratory setting?

In the absence of a truly comparative study of the phenomenon, one can speculate that the laboratory situation is itself out of context. In addition, the cues provided are frequently partial and artificial, the behaviors of other participants are unfamiliar or even contradictory to the subject's past experience, and the immediate setting so new and strange that the subject becomes confused, unsure about what to do, and hence highly susceptible to what Orne (1962, 1973) has called the demand characteristics of the experiment.

The best example of this dynamic is found in the previously cited experiments by Milgram in which the cry for help comes over an intercom from an unseen person in an unseen room, and the experimenter instructs the subject to disregard the desperate call. The effect of such an out of context experience is to increase anxiety and at the same time to reduce the cues that normally elicit and guide behavior. Altruistic conduct is highly unlikely under such circumstances. As a situation approaches the limits of the human condition, humane responses are hardly to be expected.

Once again, this is not to imply that laboratory experiments are *sui generis* ecologically invalid. They are neither more nor less so than those in any other setting. As specified in our definition of ecological validity, the critical issue is that of the purpose for which the experimental setting is employed, the environment it is presumed to represent. Thus Milgram deliberately sought to establish an impersonal situation in which the subject would be overwhelmed by the power and paraphernalia of science and technology. By drawing on institutions and roles carrying authority in the society at large, he created a connection between the laboratory and the outside world, thus giving the laboratory experience meaning in a larger context. The result was a small-scale analogue to the situation of the individual confronted with a far more ominous authority— the delegated representative of a police state. To achieve a similar connection between a contrived experiment and social reality, Zimbardo and his colleagues moved out of the laboratory to establish a simulated prison and its constituent roles in a college dormitory.

From these examples there emerges a principle that guides the successful use of the laboratory as an ecologically valid setting for research on human behavior and development.

PROPOSITION G
The significance of the laboratory as an ecological setting employed for research on human behavior is determined by how the laboratory situation is perceived by the subjects, and by the roles, activities, and relations activated by those perceptions. Hence the laboratory becomes an ecologically valid setting for

human studies only when the two following conditions are met: the psychological and social meaning of the laboratory experience to the subject is investigated and becomes known to the researcher, and the subjective meaning of the laboratory situation corresponds to the environmental experience to which the investigator wishes to generalize.

This proposition represents simply the application of the criterion of ecological validity (definition 8) to the laboratory setting. It deserves a separate statement because it makes explicit certain assumptions and imposes requirements that are not typically heeded in laboratory studies of developmental processes and outcomes. The first such assumption is that the subject's behavior in the laboratory is a function of his definition of the situation. Second, this definition invariably has a social aspect: the laboratory is perceived as a social situation that takes its meaning from the subject's experience in the other social settings in his life.

These assumptions, in turn, impose two requirements on the researcher. First, he is obligated to investigate the psychological and social meaning of the laboratory situation for the subjects. The methods for obtaining such information may vary; the meaning may be inferred from the subject's behavior, probed in follow-up inquiry, and so on. As we have seen, a particularly powerful strategy in this regard is the comparison of behavior by the same or similar subjects in different settings. Then, in light of the evidence about how the laboratory situation was perceived by the research subjects, the investigator must evaluate the extent to which the findings can be generalized to other settings.

It will be recalled that our definition of a setting, as set forth in chapter 2, encompasses not only the immediate location of the person under study but also physical aspects of the surroundings—objects, equipment, or any other features that can affect the course of events. Some of these features, while characteristic of a particular setting, can also be moved into other locales. Taking advantage of this fact, scientific investigators have frequently introduced laboratory methods into the home, the classroom, and other natural environments, with the purpose of benefiting from the greater objectivity and control that the laboratory methods were designed to achieve.

There exists, however, the danger of a creating an *ecological dis-*

tortion by injecting into a natural situation elements that are unfamiliar and hence disorienting and disruptive of the patterns of activity and relations that normally occur in the setting. With respect to studies of parent-child interaction in the home, we have already noted the constraining effect of the researcher's instructions to the parent to remain expressionless or to refrain from speaking except when the child speaks first. We have also noted Seaver's conclusion that the failure of previous investigators to obtain significant teacher expectancy effects was attributable to the artificial and unusual nature of the experimental manipulations employed. Thus the transfer of laboratory procedures into the field can lead to the elimination of conditions normally present or to the introduction of extraneous elements. Such changes represent, on the one hand, an impoverishment of available cues and, on the other, a contamination of the familiar context. It is not unreasonable to expect that, in terms of their effect on the participants, these factors can combine to create ambiguity in the situation, generate feelings of uncertainty, and thereby result in insecurity, anxiety reactions, and impairment of performance. As in the laboratory setting, such feelings of distress and intimidation are most likely to be experienced by children, members of ethnic minorities, low income populations, and other groups with limited education or familiarity with the world of scientific research.

Such considerations extend to the use of psychological techniques in general. A number of social scientists (Labov, 1967; Mercer, 1971; Sroufe, 1970; Tulkin, 1972) have called into question the general use of standardized tests of intelligence and achievement in schools, social agencies, and the courts as a basis for decisions affecting a person's life. Labov (1970) has argued that the seemingly retarded patterns of speech and general behavior often found when disadvantaged children are examined in school settings represent reactions to an alien environment. In support of this thesis, he presents evidence that the same children, when observed and interviewed in their homes under relaxed circumstances, can become fluent and effective communicators of their ideas.

An ironic twist on this same theme emerges from a study by Seitz et al. (1975) of economically deprived preschool children who were tested on the Peabody Picture Vocabulary Test either in their own homes or in an office at a school or Head Start center. All the children were black and lived within a low income inner city neighborhood. They were tested by a white, middle class examiner. Half the children had been enrolled for five months in a Head Start program. A comparable group without preschool experience was identified by consulting Head Start waiting lists and canvassing Head Start-eligible families.

As was expected, the Head Start children scored significantly higher than the comparison group. The authors interpret this result as "more likely a reflection of changes in motivational factors than of changes in formal cognitive abilities" (p. 482). The data on the effects of test location yielded a somewhat surprising finding: whereas children with Head Start experience did equally well no matter where the test was administered, the non-Head Start children scored significantly lower when tested at home than in an office at the school or at the preschool center. Puzzled by this result, the investigators conducted an inquiry after the fact and reported the following:

Initially the finding . . . seemed counterintuitive that any child would perform more poorly in his own home than in the unfamiliar office. Observations made by the examiner in the present study, however, suggested that the introduction of a middle-class examiner into the child's home created an unusual situation for the child. For example, the children were dressed as for a special occasion rather than in casual playclothes. Mothers generally remained close to their children and monitored their performance, either by busying themselves in a nearby room or by coming into the testing room frequently on errands. They also showed concern afterwards, asking how their child had performed, and they communicated the unusualness of the situation, as in instructing the children, "Don't touch that. I've just cleaned up in here." A mother's attitude thus may have caused the situation to become permeated with tension and anxiety for the child. This anxiety in turn may have been important in attenuating the child's test performance. (Pp. 484–485)

This interpretation of an unexpected finding supports the notion that ecological distortions introduced by established scientific procedures are not limited to the laboratory. It would appear that proposition G, originally formulated to meet the special ecological vulnerabilities of the laboratory, actually applies to any setting employed for research on human behavior. Since it is easy to assume that an inquiry conducted in a natural environment is ecologically valid, it may be prudent to restate proposition G in a more general form applicable to laboratory, home, school, workplace, or any other locale into which the inquisitive investigator of human behavior may wander.

PROPOSITION G'
A setting becomes ecologically valid for research on human
behavior and development only when the following two conditions are met: the psychological and social meaning of the
subject's experience in the setting is investigated and becomes
known to the researcher, and the subjective meaning of the
research situation corresponds to the environmental experience to
which the investigator wishes to generalize.

It remains true, however, that the requirements of this proposition are more likely to be violated in the laboratory than in natural
environments. The social phenomenology of the research setting is
seldom examined in laboratory studies, particularly in developmental psychology. Indeed the effort has been in the opposite direction: to exclude the subjective from the domain of rigorous scientific
inquiry. The stricture has seldom been challenged in experimental
psychology, probably because so much of its work has been done
with animals. But once again, a research model that may be reasonably adequate for the study of behavior and development in
subhuman species is insufficient for the human case.

The importance of the social perception of a research setting was
recognized earlier by sociologists than by psychologists. From
the very beginnings of their discipline, the former were more
oriented toward studying events in their social context. It was the
Chicago school of Cooley (1902), G. H. Mead (1934), and, in
particular, W. I. Thomas (Thomas, 1927; Thomas and Thomas,
1928; Thomas and Znaniecki, 1927) who stressed the importance
of the person's subjective view—in Thomas's language the *definition
of the situation*—as a major determinant of action.

In psychology, the growth of interest in perceived as against
objective reality took a somewhat different course; the emphasis
was almost entirely restricted to cognitive processes and all but
excluded the social realm. The development had its beginnings in
the work of European psychologists on the phenomenology of visual
perception (Husserl, 1950; Katz, 1911, 1930; Koffka, 1935; Köhler,
1929, 1938; Wertheimer, 1912). A significant expansion of subjective analysis to the cognitive sphere was carried out by Piaget
through his theoretical conceptions and his advocacy of the "mé-
thode clinique" to probe the processes occurring in the mind of
the child (1962), but with no reference to social context. Lewin did
extend phenomenological analysis beyond the individual to the

study of the environment with his concept of the "psychological field" (1935), but his intricate diagrams of the person's life space remained a true topology, lacking either psychological substance or social structure (Bronfenbrenner, 1951). It was not until the late 1940s, when MacLeod (1947) published his classic paper "The phenomenological approach to social psychology," that the importance of studying the person's subjective view of social reality was given explicit recognition.[5] MacLeod emphasized the need to answer the question: "What is 'there' for the individual . . . What is the social structure of the world he is living in?" (p. 204). But in general MacLeod's injunctions remained unheeded: experimental studies of human development continued to be overwhelmingly behavioristic in theory and method and thereby uninterested in and ignorant of the meaning of the research experience for the subject. The omission is critical in research on human beings, for, as Mead pointed out (pp. 304–355), it is precisely in our capacity to attribute meaning to stimuli that we differ most from subhuman species.

The reasons for resistance to a phenomenological approach in the scientific study of human functioning lie, I suspect, in the history of psychology as a discipline. Empirical psychology has based its research paradigms on physics rather than on the natural sciences. The aim of science was seen as the establishment of universal propositions through the observation of *objective* phenomena under maximally controlled conditions. The need to eliminate the "personal equation" in early studies in astronomy led to the exclusion of subjective experience as an extraneous variable. Given these requirements, investigations conducted in real-life settings have been regarded as scientifically suspect and unsuited to the task, since such settings are clearly subject to influences from manifold sources, vary considerably over space and time, and are therefore likely to be highly particular. As a result, so the argument has run, naturalistic studies are far more likely to produce situation-specific findings than to reveal general developmental laws. One is in a much better position to discover such universal principles in the laboratory, where extraneous variables can be excluded and conditions kept almost identical from one experiment to the next.

This position has been elegantly expounded by Weisz (1978). Accepting a model from physics as the scientific ideal, he argues:

If I wished to discover principles that govern the falling of objects, I might station myself beneath a tree in autumn and observe falling leaves. Unperturbed by the influence of variations in wind direction and velocity or leaf shape and size, I might determine that the central tendency of the falling leaves was to follow a southerly path 10° off the vertical. Subsequent experiments beneath other trees with different prevailing winds might gradually lead me closer to the truth about how "unadulterated gravity" operates, but this truth might have been quickly discovered in one simple, though ecologically invalid, experiment conducted in a vacuum chamber (a nonenduring, not naturally occurring setting).

Similarly, in determining (or testing a hypothesis as to), say, whether the concept of identity or of transitivity has developmental precedence, observations in various natural settings of children's behavior with various naturally occurring tasks involving the two concepts (but, inevitably, differing intellectual demands as well) might leave my view of the principle under study obstructed by factors in the social situation and the games themselves, which are of no particular interest to me at the time. In this case, a good deal might be learned rather quickly by means of an artificial setting with the social situation carefully structured and a contrived, short-lived task with demands on such intellectual processes as attention and memory carefully minimized. (P. 6)

The difficulty with this line of reasoning is its assumption that physical and psychological objects and environments are equivalent. Although the paradigms of physics have high status in the scientific psychological establishment, they unfortunately provide a false foundation for the discipline because they often are ecologically invalid. It is of course entirely valid scientifically to place a falling object in a physical vacuum to study the laws of its behavior under optimally controlled conditions; but it is *not* valid, for similar purposes, to place a *person* in a social vacuum for the simple reason that the human being cannot function effectively under such conditions. It is like taking a fish out of water; the organism simply cannot survive for very long. And since human beings, like all living creatures, have strong survival mechanisms, the first thing a person does in such circumstances is to fill the vacuum with social meaning.

Furthermore, there is a critical difference between physical objects and human beings: physical objects cannot, and human beings invariably do, have perceptions, feelings, expectations, and intentions with respect to the situations in which they are located. And, once such processes are operative, the outcomes for people are determined by Thomas's principle: "If men define situations as real, they are real in their consequences" (Thomas and Thomas, 1928, p. 572).

The scientific moral is clear. If results are to be validly interpreted, experiments on human behavior and development cannot be conducted in a social vacuum. It is therefore necessary to anchor them in social reality in such a way that this reality is perceived by the research subjects in the manner intended by the investigator, and these perceptions can be assessed and verified as part of the experimental procedure. These are of course the requirements specified in propositions G and G'.

The effect of proposition G' is to give the laboratory the same status as any other ecological setting, which means that the social significance of the setting for the research subjects has to be established before their behavior can be understood and its implications for development determined. Even after this task has been accomplished, the problem that Weisz sees the laboratory as solving— what he calls "transcontextual validity" or "veridicality of principles across contexts" (p. 2) remains.[6] For the fact that a particular process occurs under controlled conditions in the laboratory does not necessarily mean that the same process operates in the same way in other settings. The critical issue is whether the several settings, including the laboratory, have equivalent psychological and social meaning to the participants.

One can also question whether establishing transcontextual validity is, as Weisz seems to imply, the ultimate goal in the scientific study of human development, which he defines as one of finding psychological "universals," developmental principles "that can be shown to hold good across physical and cultural setting, time, or cohort" (p. 2). A distinctive property of human beings, however, appears to be precisely their capacity to adapt—to respond differentially to diverse physical and cultural settings. Given the ecologically dependent character of behavior and development in humans, processes that are invariant across contexts are likely to be few in number and fairly close to the physiological level. What behavioral scientists should be seeking, therefore, are not primarily these universals but rather the laws of invariance at the next higher level— principles that describe how developmental processes are mediated by the general properties of settings and of more remote aspects of the ecological environment.

Paradoxically, it is in this second area that the laboratory can make a unique contribution. By serving as a controlled and contrasting setting, it can illuminate the distinctive properties of the more enduring contexts in which human beings live. For this reason it is important to replicate, or at least simulate, in the laboratory natural or contrived experiments, that have occurred in real-life settings, and vice versa.

Weisz (1978) and Parke (1979) view research methods as lying along what Parke refers to as "a continuum of naturalness." The naturalistic end of the continuum offers, in his view, the advantages of greater ecological validity, whereas the opposite end provides more opportunity for introducing and maintaining strict experimental controls. Both Weisz and Parke argue that, under these circumstances, the scientific strategy of choice is one that uses methods at different points along the continuum to establish the generality or—in Weisz's terminology, the transcontextual validity— of a given hypothesis.

From an ecological perspective, there is a problem with this formulation. For reasons indicated earlier, conducting research in a natural setting does not confer an automatic guarantee of ecological validity. Furthermore, laboratory research is not always ecologically invalid. As we have seen, laboratory situations can possess psychological and social meanings that are as real and compelling as those in any real-life setting. The difficulty is that in present scientific practice these often remain unexamined. To advocate the complementary use of laboratory and field methods without recognizing these methodological and substantive complexities is not only to ignore fundamental problems of scientific validity but also to overlook the possibility of combining the two approaches in a way that takes advantage of the distinctive ecological properties of each.

Both Weisz and Parke suggest by their arguments and examples that the purpose of conducting investigations in different settings is to establish the generality of a particular hypothesis. No recognition is given to the scientific importance of investigating the ways people are differentially affected by the environments in which they find themselves or—in more theoretical terms—how psychological and social processes are mediated by the context in which they occur. If the legitimacy and priority of such questions were recognized, they would argue for some modification in the prevailing substantive focus and division of labor among researchers on human development. As it is, the primary purpose of conducting studies in different settings is to establish the generality of a given process across settings. Thus the differential properties of these settings are not considered particularly important; what counts is that the process in question be demonstrated as occurring in each. Nor does it matter very much whether the relevant studies are conducted by the same or by different investigators, working independently or in collaboration. Each can make a contribution, and there need be no confrontation between them. To quote Weisz's conclusion, "It is essential that proponents of these two complementary pursuits engage in their work with mutual respect, and that neither seek to impose gratuitous limits on the scope of developmental inquiry" (p. 10). Weisz's statement actually was made with reference to the traditional distinction between basic and applied research. As previously noted, he regards the former as generally better served by the laboratory and the latter by studies in natural settings. Although the two can complement each other, they are seen as "oriented toward different values." The aim of the first is the "quest for firm scientific principles," whereas the second seeks to enhance "the immediate well-being of contemporary society" (p. 10).

So long as basic science is defined as the search for universal processes that are invariant across contexts—and this is the position taken explicitly by Weisz and implicitly by Parke and his colleagues —Weisz's conclusions logically follow. But if the study of human development includes the pursuit of principles that govern the way in which processes of behavior and development are instigated and altered by the environments in which they occur, then more integrative conceptions and strategies are required. The conceptual systems of developmental science must encompass the general properties of settings as well as of persons and behaviors; they must also allow for interaction between person and situation in affecting behavioral outcomes. This requirement means that no research can be conducted without some analysis of the setting in which it is carried out or an explicit consideration of the relation to other ecological contexts and generalizability of the findings. This, in turn, implies some scientific shortcomings to research that is restricted to a single setting and is pursued within a theoretical framework that makes no allowance for other ecological parameters. The same line of reasoning places a scientific premium on work that draws on data from more than one setting, with close collaboration and exchange of ideas between investigators working in different settings.

As I have argued elsewhere (Bronfenbrenner, 1974a), a conception of the discipline as one primarily concerned with *development-in-context* implies not merely a complementary relation but a functional integration between scientific and social policy concerns. Public policy questions are relevant for basic science primarily because they can alert the researcher to aspects of the immediate and, especially, the more remote environment that affect developmental processes and outcomes.

The functional interplay between science and public policy of the type here envisioned implies a closer, more integrated, and more productive relation between laboratory and field research. The present pattern is one of parallel, functionally independent activities which, if they converge at all, do so by seeking similar results attained by different methods to support hypotheses about psychological processes presumed to be invariant across social contexts. If my analysis of available data and their theoretical implications is correct, it argues that further advance in the science of human development requires addressing a higher order of invariant relations, those that describe developmental processes as a function of systematic properties of the ecological contexts in which they take place. To accomplish this challenging task, investigators must look upon the laboratory, or any other research setting, not merely as a place in which to conduct research but as a central object of study that invariably affects the processes being observed and hence can shed light on the environmental forces that steer the course of human development.

7.

Children's Institutions as Contexts of Human Development

Besides the family home, the only setting that serves as a comprehensive context for human development from the early years onward is the children's institution. From an ecological perspective, the existence of such a context is important because it provides an opportunity to investigate the impact of a contrasting primary setting on the course of development through childhood, adolescence, and sometimes beyond, into the middle years and old age.

Unfortunately for our purposes, most investigations of development in institutions have, in keeping with the characteristic focus of the traditional model, concentrated on psychological outcomes for the individual with almost no attention to the structure of the immediate environment, or in our terms, of the microsystem in which the individual is embedded. Little information is provided about the complex of activities, roles, and relations that characterizes the institutional setting and differentiates it from the more common developmental context of family and home.

There are, however, a few notable exceptions. Chief among them is the classic but highly controversial comparative study published in the 1940s by the psychoanalytically oriented psychiatrist René Spitz (1945, 1946a, 1946b). Spitz's work has been, and continues to be, subjected to considerable criticism on both methodological and substantive grounds. The more severe among his critics (Clarke and Clarke, 1976; O'Connor, 1956, 1968; Orlansky, 1949; Pinneau, 1955) have taken the position that his hypotheses were ambiguously stated, his research design confounded, his data not trustworthy, his analyses inadequate, his results never replicated, and his findings subject to more parsimonious alternative interpretations; in short, they view his case as demolished and his thesis as a dead issue no longer worthy of serious consideration.

132

I take a contrary position on each of these scores. A painstaking examination of Spitz's original work, related studies, and the now-extensive commentary by Spitz's critics and defenders, has led me to conclude that as an investigator of development-in-context he was a man ahead of his time and that his experiment, while not without serious flaws, represents an early prototype of an ecological model. His analysis of the distinctive properties of the children's institution as a developmental context was essentially in terms of what I have called molar activities, roles, and, in particular, joint activities and primary dyads engaged in both by children and their caretakers. In addition, I contend that both Spitz's hypotheses and his research design were more precise than his critics recognized and that his results provide support and clarification of hypotheses regarding the central process in the ecology of human development —"the progressive, mutual accommodation between an active, growing human being and the changing properties of the immediate settings in which the developing person lives" (definition 1).

To grasp the significance of Spitz's work, it is necessary to understand the origin and nature of the conflict that it engendered, a conflict that continues to the present day. It has become a conflict not about facts but about their interpretation. There is substantial evidence, summarized by me (Bronfenbrenner, 1968) and more recently updated and reviewed by Clarke and Clarke (1976), that *under certain conditions* institutionalization of children at an early age results in impairment of psychological function and development. The source of disagreement is in defining the nature of these critical conditions. One group of experts, primarily psychoanalytically oriented (Ainsworth, 1962; Fraiberg, 1977; Heinicke, 1956; Heinicke and Westheimer, 1965), follow Bowlby (1951) in contending that the critical factor is maternal deprivation, the absence or severing of a mother-child bond. Their antagonists, by and large trained in the tradition of laboratory research emphasizing physical stimuli (Casler, 1968; Dennis, 1960; Dennis and Najarian, 1957; Dennis and Sayegh, 1965; O'Connor, 1956, 1968; Orlansky, 1949; Pinneau, 1955), argue that the observed effects are not specific to the mother-child relationship at all but derive from a general impoverishment of environmental stimulation in the institutional setting.

From the very beginning, Spitz's research has occupied a central position in this debate. Spitz conducted an observational and psychometric study of the development during the first year of life of children in four environments in two different Western Hemisphere countries. Two of these environments (one in each country) were institutional, and two involved infants from a similar cultural background raised in their own homes. Whereas with few exceptions the infants in one of the institutions (nursery)—and in both the control groups—showed normal development throughout the year (as measured by the Hetzer-Wolf baby test), those in the foundling home exhibited a marked drop in developmental quotient from 124 to 72. During this same period the figures for the corresponding control group of children raised in their own homes were 107 and 108. By the end of the second year, the developmental quotient of the foundling home infants had fallen to 45. In addition to severe developmental retardation, these children exhibited high susceptibilty to infection as well as markedly abnormal behavior ranging from extreme anxiety and bizarre stereotyped movements to profound stupor.

According to Spitz, the progressive retardation of children in the foundling home could not be attributed to poor nutrition or medical care (which he describes as comparable in the two institutions) or to the social background of the mothers, which was actually better in the case of the foundling home children. Major differences favoring the nursery, however, did exist between the two environments. In the nursery, each child received the full-time care of its own mother or, in exceptional instances, of a mother substitute. When not being cared for by the mother, the child lay in an individual cubicle enclosed in glass permitting full range of vision and, at six months of age, was transferred to rooms containing four or five cots in each. In the foundling home, infants were cared for by their own mothers only until weaning, which occurred at the beginning of the fourth month; thereafter all care was in the hands of nurses, with a ratio of eight children per caretaker. Until fifteen to eighteen months of age the children lay in cots with bed sheets hung over the railings so that the child was "effectively screened from the world" (1945, p. 62). Moreover, "probably owing to the lack of stimulation the babies lie supine in their cots for many months and a hollow is worn into their mattresses . . . this hollow confines their activity to such a degree that they are effectively prevented from turning in any direction" (p. 63). Furthermore, at least during the early part of the study, the children in the foundling home had fewer toys than their age-mates in the nursery.

In evaluating the factors accounting for the progressive deterioration among the foundling home infants, Spitz concluded that maternal deprivation played the crucial role, and it is this inter-

pretation that has been repeatedly criticized beginning with Orlansky (1949) and continuing to recent times (Clarke and Clarke,
1976). The principal charge has been Spitz's "failure to distinguish
separation from other possible (and more probable) causes of
retardation" (Clarke and Clarke p. 9), notably "the unstimulating
nature of the institution environment" (O'Connor, 1956, p. 184).

Spitz did in fact give explicit consideration to the possibility that
the progressive deterioration exhibited by the foundling home infants was "due to other factors . . . such as the perceptual and motor
deprivations from which they suffer" (1945, p. 66). Nevertheless,
he accorded primary importance to the mother-child relationship:
"It is true that the children in the Foundling Home are condemned
to solitary confinement in their cots. But we do not think that it is
the lack of perceptual stimulation *in general* that counts in their
deprivation. We believe that they suffer because their perceptual
world is emptied of human partners, that their isolation cuts them
off from any stimulation by any persons who could signify mother-
representatives for the child at this age" (p. 68).

In support of his conclusion Spitz offers several lines of evidence
and argument. First, he calls attention to the point at which foundling home babies began to fall below their age-mates in nursery.
This change occurred between the fourth and fifth months, that is,
shortly after they had been weaned and turned over to the care of
a nurse responsible for seven other infants. "The inference," writes
Spitz, "is obvious. As soon as the babies in Foundling Home are
weaned the modest human contacts which they have had during
nursing at the breast, stop and their development falls below
normal" (p. 66). This interruption of human contact, argues Spitz,
is critical since an inanimate perceptual stimulus can only be of
minor importance to a child under twelve months of age without
"the intervention of a human partner, i.e., by the mother or her
substitute" Spitz then develops his thesis as follows:

A progressive development of emotional interchange with the other provides the child with perceptive experiences of its environment. The child
learns to grasp by nursing at the mother's breast and by combining the
emotional satisfaction of that experience with tactile perceptions. He
learns to distinguish animate objects from inanimate ones by the spectacle provided by his mother's face in situations fraught with emotional
satisfaction. The interchange between mother and child is loaded with
emotional factors and it is in this interchange that the child learns to
play. He becomes acquainted with his surroundings through the mother's
carrying him around; through her help he learns security in locomotion

as well as in every other respect. This security is reinforced by her being at his beck and call. In these emotional relations with the mother the child is introduced to learning, and later to imitation. We have previously mentioned that the motherless children in Foundling Home are unable to speak, to feed themselves, or to acquire habits of cleanliness; it is the security provided by the mother in the field of locomotion, the emotional bait offered by the mother calling her child, that "teaches" him to walk. When this is lacking, even children two to three years old cannot walk. (P. 68)

Spitz's analysis of how developmental processes are mediated by the mother-child relationship corresponds with my own formulation in hypothesis 7 of the role of a primary dyad in facilitating learning and development, where the key factor was identified as "the participation of the developing person in progressively more complex patterns of reciprocal activity with someone with whom that person had developed a strong and enduring emotional attachment."

The fact that Spitz's interpretation is consistent with a general principle derived from other data does not, of course, establish its validity in regard to his own findings. It could still be argued that the observed effects were due to "perceptual stimulation in general," and that separation from the mother was not an essential element. To demonstrate that the latter plays a central role, Spitz would have had to show that such separation can bring about extreme reactions even in the absence of general deprivation of the kind encountered in the foundling home and, conversely, that in the absence of a severed mother-infant bond the observed syndrome does not occur.

Although it is not generally acknowledged, Spitz did undertake such a project, described in a subsequent report (1946b). He investigated 19 cases of profound emotional disturbance that did occur among the 123 infants in the nursery, where conditions of general stimulus deprivation did not prevail—the babies were kept in cots permitting free visibility and locomotion and were well provided with toys. Their syndrome, similar to that observed in the foundling home, was one of severe withdrawal and emotional disruption.

While unrelated to sex, race (both black and white children were included), or developmental level prior to onset (as measured by the Hetzer-Wolf Test), the syndrome in the nursery (which Spitz refers to as "anaclitic depression") appeared only in children within a delimited age range—from six to eleven months. Spitz notes that in all cases, the mother was removed from the child between the sixth and eighth month, and remained absent for a period of three months. No child whose mother was *not* removed developed the syndrome. At the same time, "Not all children whose mothers were removed developed the same syndrome. Hence, mother separation is a necessary, but not a sufficient cause for the development of the syndrome" (p. 320).

Spitz then developed a hypothesis about the nature of the "sufficient cause" and went on to report some data that he viewed as providing indirect confirmation for his hypothesis. He obtained ratings from staff members of the quality of the mother-child relationship prior to separation. Among the twenty-six mothers rated as having a "good" relationship with their infant, there were seventeen cases of severe and four cases of mild depression. Among thirty-eight mothers with a "bad" relationship, there were only eleven cases of depression, all of them mild. The value of X^2 for the two-by-three table (not given by Spitz) was statistically significant. In other words, depression was both more frequent and more severe in cases of a good mother-child relationship.

Not content with correlational evidence, Spitz then undertook an experimental manipulation. It consisted in returning the mother to the child in the three instances in which this proved possible.

The change in the children's observable behavior was dramatic. They suddenly were friendly, gay, approachable. The withdrawal, the disinterest, the rejection of the outside world, the sadness, disappeared as if by magic. But over and beyond these changes most striking was the jump in the developmental quotient, within a period of twelve hours after the mother's return; in some cases, as much as 36.6 per cent higher than the previous measurement [taken shortly before the mother returned].

Thus one would assume that if adequate therapeutic measures are taken, the process is curable with extreme rapidity and the prognosis is good. (P. 330)

Spitz contends that the same therapeutic effect can be achieved by providing a mother substitute and points to the contrasting course of events that took place in the nursery. When separation occurred in that setting, a different practice was followed, which Spitz credits with having prevented developmental decline. "The separation of the infants from their mothers takes place in Nursery between the sixth and the ninth month. Another of the inmates is then assigned to the care of the motherless child" (p. 335). He goes on to emphasize, however, that the provision of a mother substitute is not a sufficient condition. "If active attempts at substitution are to be initiated through social contact, locomotion is a necessary prerequisite for such an attempt. In institutionalized children, both the opportunity to reestablish . . . relations through social contact, and the opportunity for locomotion, are severely handicapped" (p. 334).

The results of Spitz's secondary analysis and experimental manipulations may be conceptualized as a complementary thesis and antithesis. The thesis asserts that the most traumatic effects of maternal separation occur at a point shortly after six months of age; both younger and older children are likely to show less severe reactions. This thesis of developmental retardation is countered by a therapeutic antithesis: the regressive trend can be prevented and reversed by providing the child with a parent substitute in an environment that allows for locomotion and spontaneous activity on the part of the child.

Findings in support of Spitz's two-sided hypothesis come from a variety of sources. As early as 1932, Bayley in a systematic study of the crying of infants during mental and physical tests reported that crying in response to a strange person increased markedly after six months of age, reaching a maximum at about ten months and accounting for a "large proportion" of all sources of crying (such as fatigue, hunger, postural discomfort, and so on) during the second half-year. Bridges' (1932) finding that the capacity of the infant to distinguish one individual from another begins at about six months of age is supported by studies of the development of the smiling response (Ahrens, 1954; Ambrose, 1961; Gewirtz, 1965; Spitz, 1946c). In an experimental study of this phenomenon, Morgan and Ricciuti (1965) conclude, "The results give general support to the common observation that infants are less likely to smile and more likely to show indications of fear of strangers after six or seven months than before" (p. 18).

Even more direct evidence comes from research on children's reactions to the experience of separation. Thus Yarrow (1956, 1961, 1964) and Yarrow and Goodwin (1963) showed that an individualized relationship with the mother develops gradually, that the relationship becomes established at about six months of age, and that separated infants show the most disturbances "*after* the development of a focused relationship with the mother" (1964, p. 122). The several papers by Schaffer and his colleagues (Schaffer, 1958, 1963, 1965; Schaffer and Callender, 1959; Schaffer and Emerson, 1964) all confirm the thesis. Three sets of findings seem particularly relevant. Schaffer and Emerson, using a measure of social attachment based on the mother's report of her child's protest (usually crying) to seven everyday separation situations (such as being put down after being held, left in the cot at night, left alone in the buggy outside shops, left with other people), plotted the development of the child's dependency on the mother during the first eighteen months of life (figure p. 23). The curve for specific (as distinguished from indiscriminate) attachment begins to rise from basal level at about six months of age—just as the curve for indiscriminate attachment begins to drop—and reaches its maximum at about ten months. In addition, the authors find that individual differences in the strength of the infant's attachment are correlated significantly and positively with two aspects of the mother's behavior: her responsiveness to the child's crying and the frequency with which she stimulates or interacts with the infant. In short, dependency increases as a direct function of stimulation. The significance of the strength of attachment for more complete and extended separations is indicated by results of an earlier study of infants' reactions to separation during hospitalization (Schaffer and Callender, 1959). To quote from the authors' summary,

Under standardized conditions of observation and analysis of data, seventy-six infants less than twelve months of age were observed while undergoing short-term hospitalization . . . two main syndromes emerged, each linked to a certain age range, with a dividing point at approximately seven months. Those above this age showed essentially the same type of behavior as that described in other studies of the preschool group, namely, considerable upset when admitted to hospital and a period of disturbance after return from home, both centering around the need for the physical presence of the mother. In the group less than seven months of age, on the other hand, separation from the mother evoked no observable disturbance and, instead, an immediate adjustment to the new environment and the people in it was found . . .

On the basis of the data presented it is suggested that the critical period, when separation from the mother is experienced as a traumatic event, does not commence until after the middle of the first year of life. (P. 539)

A third study by Schaffer (1965) provides additional confirmation for the hypothesis of a critical period while strengthening the weakest empirical link in Spitz's chain of evidence. The latter's contention that the effects of institutional deprivation could be prevented or reversed was based on only a few case studies. Schaffer presents more systematic and reliable support for this thesis in an experiment involving two groups of twenty-two infants separated from their mothers during the first seven months of life and subjected to differing degrees of environmental stimulation in the institutions in which they were placed. Differences occurred in opportunities for locomotion, caretaker-child ratios (1:6 versus 1:2.5), and levels of attention from caretakers as measured by a time-sampling technique. Length of stay in each institution ranged from two to nine months. To measure developmental changes, the Cattell Infant Intelligence Test was administered within seven days of admission, three days prior to discharge, about two weeks after return home, and three months afterward. Despite the fact that both groups were separated from the mother, only the infants in the more deprived setting showed a significantly lower developmental quotient (85 versus 95) during the period of institutionalization. After return home the same group achieved a significant rise, whereas the controls remained constant. As Schaffer points out, this pattern of results is consistent with the conclusion that the critical factor about institutionalization during the first half year of life is not separation from the mother but restriction of opportunity for caretaker-infant interaction and general activity by the infant.

Findings consistent with Spitz's therapeutic hypothesis using a sample of older infants appear in a report by Provence and Lipton (1962). A group of fourteen infants who had shown effects of institutional deprivation during the first year of life were placed in foster homes at between nine and twenty-nine months of age. The ensuing developmental changes are described by the authors as follows:

The infants who had shown the retardation and disturbed development in the first twelve to eighteen months described in the major part of this report made dramatic gains when given the benefit of good maternal care and family life. In many aspects of their development they looked sufficiently improved that they were not markedly different from their peers on superficial observation and casual contact.

However, when one looked more closely one could recognize certain characteristics which appear to be related to the absence of adequate maternal care with all that this implies. The areas in which there were residual impairments of mild to severe degree were in their capacity for forming emotional relationships, in aspects of control and modulation of impulse, and in areas of thinking and learning that reflect multiple adaptive and defensive capacities and the development of flexibility in thought and action. A lessened capacity for the enjoyment and elaboration of play and an impairment of imagination were also evident. One missed in them the characteristic of the healthy family child of richness and originality in the personality in which one perpetually discovers or catches glimpses of some new facet. (P. 158)

Regrettably, the conclusions about residual effects of institutionalization were not based on a systematic comparison with a matched control group of children raised from birth in family settings.

Findings consistent with Spitz's hypothesis regarding effects of institutionalization on older children, and carrying broader implications for public policy, came from a naturalistic experiment reported by Prugh and his associates (1953). These investigators took advantage of a planned change in hospital practice to conduct a comparative study of the reaction of children (and their parents) to two contrasting modes of ward operation. There were one hundred youngsters in each group, ranging in age from two to twelve. All had been admitted to the hospital for short-term care involving medical methods of diagnosis and treatment, as distinguished from surgery. The groups were matched in age, sex, diagnosis, number of previous hospitalizations, and average length of stay for the present admission (approximately one week). The control group consisted of children admitted and discharged over a four-month period prior to the introduction of the contemplated change. The young patients hospitalized during this period experienced "traditional practices of ward management" (p. 75) in which parents were restricted to weekly visiting periods of two hours each. The experimental group, admitted during the next period, could receive daily visits from parents. In addition, the parents accompanied the child on his admission to the ward, were introduced to staff, and were encouraged to participate in ward care. Other innovations included early ambulation of patients, preparation and support for potentially distressing medical procedures, and a special play program.

Significantly greater emotional distress was observed among the children in the control group not only while they were in the hospital but also during follow-up visits to the home conducted periodically throughout the year following hospitalization. Ninety-two percent of the children in the control group "exhibited reactions of a degree indicating significant difficulties in adaptation" (p. 79) compared with 68 percent for those in the special program. The intensity and duration of disturbance varied inversely with the age of the child: three months after discharge, half the children under three in the control group, as opposed to 37 percent of the experimental group, showed evidence of "severe" disturbance as manifested by such behaviors as "constant crying," "outbursts of screaming when approached by an adult," "refusal to chew food," and "loss of bowel and bladder functions." (p. 88). The corresponding figures for the six to twelve-year-old group were 27 percent and zero respectively. In general, the older the child, the greater was the positive impact of the experimental program.

The most powerful evidence for Spitz's therapeutic hypothesis comes from an experiment conducted by Skeels and his colleagues in the 1930s (Skeels, 1966; Skeels and Dye, 1939; Skeels et al., 1938). The subjects were two groups of mentally retarded, institutionalized children. The average Binet IQ of the children and of their mothers was under 70. When the children were about two years of age, thirteen of them were placed in the care of female inmates of a state institution for the mentally retarded with each child being assigned to a different ward. The control group was allowed to remain in the original, also institutional, environment— a children's orphanage. During the formal experimental period, which averaged a year and a half, the experimental group showed a mean rise in IQ of 28 points, from 64 to 92, whereas the control group dropped 26 points. Upon completion of the experiment, it became possible to place eleven of the experimental children in legal adoption. After two and a half years with their adoptive parents, this group showed a further nine-point rise to a mean of 101.

In Skeels's view, the key to the success of the experimental intervention was the relationship that developed between the child and an adult on the ward:

it must be pointed out that in the case of almost every child, some one adult (older girl or attendant) became particularly attached to him and figuratively "adopted" him. As a consequence, an intense one-to-one adult-child relationship developed, which was supplemented by the less intense but frequent interactions with the other adults in the environment. Each child has some one person with whom he was identified and who was particularly interested in him and his achievements. This highly stimulating emotional impact was observed to be the unique characteristic and one of the main contributions of the experimental setting. (1966, p. 17)

But the interpersonal relationship was not the only feature of the ward setting that contributed to the children's development. There were at least two other significant elements:

the attendants and the older girls became very fond of the children placed on their wards and took great pride in them. In fact, there was considerable competition among wards to see which one would have its "baby" walking or talking first. Not only the girls, but the attendants spent a great deal of time with "their children" playing, talking, and training them in every way. The children received constant attention and were the recipients of gifts; they were taken on excursions and were exposed to special opportunities of all kinds . . .

The spacious living rooms of the wards furnished ample space for indoor play activity. Whenever weather permitted, the children spent some time each day on the playground under the supervision of one or more older girls. Here they were able to interact with other children of similar ages. Outdoor play equipment included tricycles, swings, slides, sand boxes, etc. The children also began to attend the school kindergarten as soon as they could walk. Toddlers remained for only half the morning and 4- or 5-year-olds, the entire morning. Activities carried on in the kindergarten resembled preschool rather than the more formal type of kindergarten. (Pp. 16–17)

In the studies we have examined, certain common features are apparent. Regarding the issue of institutional deprivation, there appear to be two environmental conditions that are critical in producing debilitating effects on children. They may be stated in the form of a hypothesis.

HYPOTHESIS 15
An institutional environment is most likely to be damaging to the development of the child under the following combination of circumstances: the environment offers few possibilities for child-caretaker interaction in a variety of activities, and the physical setting restricts opportunities for locomotion and contains few objects that the child can utilize in spontaneous activity.

Another set of findings emerging from the researches we have reviewed identifies a critical period in the child's life when institutionalization has its maximally disorganizing impact.

HYPOTHESIS 16
The immediate disruptive impact of an impoverished institutional environment tends to be greatest for children who, upon entry into the institution, are separated from the mother or other parent figure in the second half year of life, when the infant's attachment to and dependence on the primary caretaker typically reaches its greatest intensity. Immediate reactions to institutionalization before or after that period tend to be less intense.

Hypothesis 16 leaves unanswered another question about the impact of institutionalization—that of long-range effects. Are reactions to institutionalization in the second half year of life not only the most intense but also the most lasting? And how long do these effects endure? In the investigations we have examined thus far, follow-up studies did not extend beyond a few months after the child was removed from the deprived environment; in some instances, sequelae were still detectable. Moreover, in none of these studies had the experimental group been exposed to the maximally depriving circumstances for longer than two years prior to removal.

It would not be surprising to find that a longer period of institutionalization has more lasting effects. (Some research bearing on these questions is reviewed below.)

These studies also point to a third and more optimistic hypothesis. It concerns the environmental conditions that can prevent or reverse the debilitating effects of institutional deprivation on children.

HYPOTHESIS 17

The developmentally retarding effects of institutionalization can be averted or reversed by placing the child in an environment that includes the following features: a physical setting that offers opportunities for locomotion and contains objects that the child can utilize in spontaneous activity, the availability of caretakers who interact with the child in a variety of activities, and the availability of a parent figure with whom the child can develop a close attachment.

The first two requirements stipulated in hypothesis 17 reverse the conditions specified in hypothesis 15 as defining a depriving environment. The third requirement, however, does not fit this pattern and thus is less self-evident. Although a person with whom the child could develop a close relationship has been present in each of the situations in which the debilitating effects of deprivation were either averted or reversed, one cannot attribute the improvement to this circumstance: in all but one of these cases, the presence of a parent figure was confounded with other therapeutic factors. The exception is found in Spitz's secondary analysis and small-scale experiment with three mother-child pairs. It is clear that there is a need for additional studies that include adequate controls.

There is little doubt that the conditions stipulated by hypothesis 17 can prevent or in substantial measure reverse the effects of institutional deprivation, but there is some question whether the recovery is complete. More broadly, it remains to be seen whether persons who have been institutionalized for some period during childhood and then discharged still show developmental effects of this experience later in life.

In Spitz's view, prolonged separation from the mother beginning in the second half year of life—after most infants have already formed a strong maternal attachment—would not only evoke a more severe reaction but would also produce effects that were more lasting than if the separation occurred earlier. In a comprehensive analysis of research on the conditions and consequences of early deprivation, I reached a contrary conclusion (Bronfenbrenner, 1968). Even though the *immediate* effects were more debilitating when maternal deprivation took place during the second half year of life, the *long-term* consequences were most severe for those institutionalized in earliest infancy. Moreover, the children separated after six months of age, despite their more traumatic initial reaction, recuperated more quickly than the latter group in response to improved circumstances. They appeared to be more susceptible to recovery through subsequent interaction with the physical and social milieu, whereas the sequelae of environmental restrictions in earliest infancy were more likely to persist into later life.

Evidence in support of this conclusion unfortunately requires the comparison of data from different studies, since no one investigation addressed the issue directly. Goldfarb (1943a, 1943b, 1955) conducted an experiment of nature that documents the effects of institutional deprivation in early infancy. His study of the residual effects of early institutionalization on boys into early adolescence is unique in its use of a control group from a comparable socioeconomic background living in similar settings at the time of assessment. For his experimental group, Goldfarb selected fifteen boys who "had entered the institution in very early infancy (mean age of four and a half months), had remained at the institution for about three years, and had then been transferred to foster homes where they had been reared up to the time of the study" (1943a, p. 107) when they averaged twelve years of age (with a range from ten to fourteen). The control group had been placed directly into foster homes at an age of approximately one year and had lived there continuously since entry. Using this design, Goldfarb was able to hold constant the experience of maternal separation in early infancy and to eliminate any effects of current participation in institutional life. He used agency records and interviews with caseworkers as a basis for rating the degree to which the child had been accepted within the family and the quality of the foster parent-child relationship. No differences between the two groups were found on either of these factors nor in other aspects of the foster home experience including the number and duration of placements, the children's facilities, and the socioeconomic and cultural background of the foster parents. There was lack of comparability, however, in two respects. Although not differing significantly in either ethnicity or amount of schooling, the mothers of the institutionalized children were somewhat superior in occupational background, a circumstance that, as Goldfarb notes, works against the hypothesis under investigation. Specifically it argues against the possibility, suggested by Clarke and Clarke (1976), that selective factors had operated in favor of the control group at the very outset. In addition, whereas both groups experienced maternal separation, this event occurred at an earlier age for the institutionalized children (mean of four and a half months) than for the controls (mean of fourteen months). In sum, Goldfarb's research design pitted a three-year experience of institutionalization beginning in the first six months of life against the experience of maternal separation at a somewhat later age but continuous residence in a family environment.

The results revealed a mean IQ at about age twelve (Wechsler Bellevue) of 72 for the institutionalized children compared with 96 for those raised in foster homes. Similar differences were found on tests of concept formation. Goldfarb also obtained observational data on the children's behavior over a period of five to seven hours: the previously institutionalized children were found to be significantly "more fearful and apprehensive . . . less responsive to sympathy or approval . . . less thoughtful in problem solution, less ambitious, less capable of sustained effort, and more prone to quit a task that is difficult" (p. 117).

Further data were obtained from a "frustration experiment" designed as follows. The child was presented with a problem for which there was only one possible solution (leaning on a table to put back a telephone receiver left off the hook). After being encouraged—and, if necessary, helped—to find this solution, the subject was asked to discover a second solution although in fact none existed. Goldfarb summarizes the results as follows:

The behavior of the institution group in a frustrating situation contrasts sharply with the foster home group. It shows a distinct tendency to be unaffected by the total situation. More of the institutional children were indifferent to success in the first place, but even more significantly more children in the institution group showed no rise in tension, were unaffected in any direction by competition and experienced no guilt or shame. If we use as a criterion of apathy the combination of absence of tension, absence of response to competition, and the absence of guilt or shame, then there are no apathetic children in the foster home group while 73% of the institution group may be characterized as apathetic.

A similar tendency is observed in a comparison of the manner of resumption. The foster home group shows a superior tendency to resumption of task after interruption. One may assume that the institution group is characterized by a lesser will to achieve, to persist and to complete a difficult task.

Finally, the institution group shows a more significant tendency to violate prohibition. In short, both groups seem to approach the problem with an equal degree of cooperation, but the institution child is emotionally more apathetic, is less concerned with success and less moved by failure or social competition, and he is more likely to disregard limitations or prohibitions. The institution child is thus adjusting on a more superficial level and is less motivated by ordinary social and human identifications. (Pp. 120–121)

It is noteworthy that the measures of developmental outcome employed by Goldfarb were not limited to scores on psychometric tests but extended to the assessment of molar activities, involving behavior in pursuit of a goal, persistence in the face of frustration, resumption of an interrupted task, deliberate adaptations to the reactions of others present in the situation, conversations, and the maintenance of ongoing patterns of interaction and enduring relationships with others.

All the above findings were obtained in a research setting removed from the child's everyday environment. It is therefore reassuring that similar results emerged from an analysis of the caseworker ratings that were based on observations in the foster homes: the following characteristics were more frequently attributed to children with prior institutional experience: "craving for affection," "lacks capacity for relationship," "restless," "hyperactive," "inability to concentrate," and "poor school achievement" (p. 123).

Since no mention is made of any effort to keep evaluators blind to the child's group status, all the observational data thus far reported may have been influenced by the observer's knowledge of the child's history. This criticism, however, is less likely to apply to information gleaned from school records. Only 20 percent of the boys with prior institutional experience were at grade level, compared with 87 percent for the foster home group. In addition, 73 percent of the former were in special classes for the retarded, whereas not one of the foster children had ever been in a special class. These differences were reflected in achievement test results, with the previously institutionalized children scoring a year below both in reading comprehension and arithmetic skills. The children with institutional backgrounds exhibited more speech defects; in particular, there was a lack of fluency and greater incidence of "errors other than those that might be classified as community errors" (p. 125).

Goldfarb offers the following general conclusions:

Following Lewin's theoretical constructs, the personality of the institution child who has been markedly deprived in infancy, is less differentiated than that of the foster child. He is more given to simple, unrefined, diffuse, unrelated or perseverative modes of adjustment. In the realm of mental organization, the more infantile plane of differentiation is reflected in the greater aimlessness of behavior, the greater preponderance of trial and error adjustment, the greater tendency to thoughtless, wasteful, non-reflective, situationally determined response. In the area of emotional organization, it is reflected in poverty of affective response, and the meagreness which is characteristic of the child's personal relationships . . . In addition, because of his isolation from adults, the institution child is severely retarded in language, has a much narrower vocabulary than his community brother, and tends to mispronounce the words he is familiar with. More significant than the specific lacks in information and language, however, is the fact that the limitation in a specific skill such as language tends to restrict the child's intellectual capacity, for it is now commonplace understanding that language and general information are active tools of thought. (Pp. 126–127)

While Goldfarb's study provides evidence of the long-range differential effects of early institutionalization versus rearing in a home environment, it does not deal directly with the question of early as opposed to late placement in an institution. Data pertinent to this issue are available, from a study by Pringle and Bossio (1958). These workers assessed the development of verbal intelligence and language, as measured by standardized tests, in samples of institutionalized children eight, eleven, and fourteen years of age. Holding age constant, the researchers investigated the independent influence on test scores of both age at admission and length of time spent at the institution. No significant differences were found for the latter factor, but the retarding effects of institutionalization were greater for children separated early and for those having no contact with their families. In addition, examination of the tables in the authors' report reveals a feature not mentioned in their analysis. Since the data are tabulated separately for three age groups (eight-, eleven- and fourteen-year-olds), it is possible to discover how differences associated with age at admission also vary with age. Significant effects are found only for the two younger age groups and never for the fourteen-year-olds. Correspondingly, the differences between means decreased with age. In other words, no deficit in verbal intelligence and language development associated with early separation from the mother could be reliably detected by the time the children were fourteen years old.

More relevant to our present concern, the degree of retardation among the institutionalized children described by Pringle and Bossio appears to be considerably milder than that reported by Goldfarb. The average IQ for the former was 90 compared with 60 for the latter, and, as I have documented elsewhere (Bronfenbrenner, 1968), the percentage of children in each sample exhibiting problem behaviors was considerably higher in Goldfarb's sample. No data are available from either study that would permit a comparison of the physical and social environments of the two institutions, but information does exist regarding one circumstance of direct significance to our inquiry. The children in Goldfarb's sample had entered the institution at an average age of four and a half months, the latest admission being at nine months. By contrast, even the so-called early-deprival group in Pringle and Bossio's sample included infants separated from the mothers at any time between birth and five years of age (no further breakdown was made). As previously noted, Pringle and Bossio found that children separated early from their mothers showed significantly lower scores in verbal IQ, but length of stay in the institution was unrelated to degree of defect.

If we are willing to make the somewhat risky assumption that children were institutionalized for more or less similar reasons in England in the middle 1950s (the locale and date of field work in Pringle and Bossio's study) as they had been in America in the early 1940s (the place and time of Goldfarb's work), we are led to conclude that, taking both sets of results into account, the critical factor in development is the age at which a child is placed in an intellectually depriving environment, with the earliest months of life being the most crucial. To be sure, there are problems with this conclusion. First, it rests on a somewhat shaky empirical foundation, buttressed at one end by only the 15 cases in Goldfarb's institutionalized group (compared with 188 in the Pringle and Bossio study). Some reassurance is provided by Goldfarb's (1943b) findings with a somewhat larger ($N = 40$) but younger sample consisting of eight-year-olds. As before, the 20 institutionalized children had been admitted before six months of age and after about two and a half years on the average, placed in foster homes. The control group of noninstitutionalized children had lived in foster homes continuously since placement also prior to six months of age. (Unlike the older sample, the two groups of eight-year-olds were separated from their mothers in the first half-year of life, so any developmental differences must be due to the influence of setting alone.) No intelligence test data were gathered, but both groups were rated on many of the same behavior items used with the older sample.

Significant setting differences in the same direction on the same traits were found for the younger sample, although neither the overall frequency of problem behavior nor the magnitude of the differences was as great. The lower frequency could of course be a function of more lenient rating standards for younger children, but the smaller size of the differences between groups might well reflect the increasing impact of early institutionalization as the boys grew older, even several years after they had been discharged.

Taken together, the results of the Goldfarb and the Pringle and Bossio studies support the conclusion that institutionalization in the first six months of life is likely to have the most enduring and damaging effects on later development, although the absence of information about conditions in the two institutions leaves open the possibility of alternative explanations. Goldfarb provides no description of the institutional environment in his original publications, but he indicates in a later report (1955) that babies were kept in individual cubicles until nine months of age and in general suggests conditions similar to those observed in Spitz's foundling home. No information whatsoever is available about the institution in Pringle and Bossio's study.

Although no definitive conclusions can be drawn, the complementary findings of the investigations reviewed here permit the formulation of a hypothesis setting forth a relation between age at institutionalization and the likelihood of long-range disruptive effects on development.

HYPOTHESIS 18

The long-range deleterious effects of a physically and socially impoverished institutional environment decrease with the age of the child upon entry. The later the child is admitted to the institution, the greater the probability of recovery from any developmental disturbance after release. The more severe and enduring effects are most likely to occur among infants institutionalized during the first six months of life, before the child is capable of developing a strong emotional attachment to a parent or other caregiver.

How long is long-range? We have seen that when children are placed in minimally stimulating institutional environments in the first six months of life, disturbance in cognitive, emotional, and social development are still detectable in early adolescence: what about young adulthood and beyond?

I have been able to find only three studies bearing on this question. Beres and Obers (1950) examined a sample of thirty-eight young adults who had been institutionalized in early infancy (typically in the first year of life) for periods ranging up to four years. Although the authors report some continuity between the early deprivation experience and later personality characteristics, they are primarily impressed by the absence of serious psychological disorder in their subjects and the gradual recovery to normal levels of functioning.

Our chief interest in this group focuses on the fact that considerable improvement to the level of satisfactory social adjustment was possible following the experience of extreme deprivation in infancy. We may again emphasize that the satisfactory adjustment in five of the seven cases did not become evident until latency or early adolescence. In their earlier years they showed varying degrees of unsatisfactory adjustment, and if our observations had been limited only to those years, we would have had to put them in one of the categories of ego maldevelopment. The implication is that the arrest of ego and superego development which characterizes the cases suffering from emotional deprivation in infancy is not an irreversible process, and that further development of ego and superego is possible . . .

Other authors have stressed the permanency of the psychological effects of extreme deprivation in infancy. Our findings are at variance with their conclusions. (Pp. 230, 232)

The same theme is echoed in Maas's study (1963) of twenty young adults who in early childhood had been evacuated from London and placed in residential nurseries for a period of at least one year. Adjustment was assessed primarily through an interview exploring different areas of personality functioning including intellectual processes, emotional control, relationships with people, and social roles. The following excerpts summarize the author's conclusions:

Though these twenty young adults may have been seriously damaged by their early childhood separation and residential nursery experiences, most of them in young adulthood give no evidence of any extreme, aberrant reactions . . .

Most of the subjects . . . do not differ significantly from a theoretically "normal" population. To this extent, this study supports the thesis of the plasticity and resiliency of the human personality. (Pp. 48, 54)

Unfortunately, in neither of these two reports were the authors able to provide information about the nature of the experiences following discharge from the institution that presumably accounted for the psychological recovery shown by the subjects. Another investigation does provide such data: thirty years after he had conducted his remarkable experiment, Skeels (1966) conducted a follow-up study of his original thirteen cases, now adults. All were found to be self-supporting, all but two had completed high school, and four had been through one or more years of college. In the control group, all were either not living or still institutionalized. Skeels concludes his report with some dollar figures on the amount of taxpayers' money expended to sustain the institutionalized group, contrasted with the productive income brought in by those who had initially been judged mentally retarded (average I.Q. of 64), had been regarded as unadoptable, but, at two years of age, were set on the road to normal development through the care of mentally deficient female patients in the wards of a state institution for the mentally deficient.

The recuperative experiences undergone by Skeels's experimental group during the course of their recovery involved virtually all the ecological principles derived thus far regarding conditions conducive to psychological growth. Once the children were placed on the wards, there were extensive opportunities for caretaker-infant interaction and for the child to engage spontaneously in a variety of activities (hypothesis 1); observational, joint activity, and primary dyads developed in due course (hypotheses 2 through 7), and third parties (hypothesis 8) entered the scene in the form of other patients and attendants. This entire process had been set in motion through a role transformation (hypothesis 9)—the female inmates were changed into child caretakers and substitute parents. This role was accorded some legitimacy, status, and power (hypotheses 10 and 11) through staff approval (hypothesis 12) and competition between the wards "to see which one would have its 'baby' walking or talking first" (p. 16).

In sum, Skeels brought about a change in every element of the microsystem—activities, roles relations, and the physical features of the setting. I shall refer to this kind of comprehensive change in all elements of the microsystem as a *setting transformation*. It represents a special case of a transforming experiment that inevitably alter the behavior of the participants and thus can affect the course of development in more powerful ways than modifications in one microsystem element at a time.

Three questions remain. First, although the three studies just examined indicate that many children subject to the debilitating effects of placement in a depriving institutional environment in the first years of life can show substantial recovery, we still do not know whether their development is truly unaffected. This could be determined only by use of a control group composed of children from comparable family backgrounds but without institutional experience. The second question relates to the quality of the institution. Thus far we have examined the developmental effects of exposure to institutional settings that offered few opportunities for physical or social stimulation. What about institutions that do provide such opportunities? And, third, do children in an institutional environment require a parent surrogate for normal development to occur?

Tizard and her colleagues (1972, 1974, 1976, 1978) assessed the cognitive development of twenty-five institutionalized children in a well-designed study including three matched control groups raised at home by their natural or adoptive parents. Both in a baseline and in a follow-up study, Tizard and Rees found "no evidence of cognitive retardation . . . in a group of four-year-old children, institutionalized since early infancy . . . In all the institutions concerned, close personal relationships between staff and children were discouraged, and the care of the children had passed through many different hands" (1974, p. 97).

The explanation for the absence of any significant disadvantage for the institutionalized children is in the next sentence: "The findings of the two studies constitute strong evidence that a good staff-child ratio, together with a generous provision of toys, books, and outings will promote an average level of cognitive development at four years in the absence of any close and/or continuous relationship with a mother substitute". As Tizard and her colleagues noted in an earlier report (1972), "The long-stay residential nurseries which we studied were very different from the grim foundling homes described by Spitz and earlier workers" (p. 339).

The work of Tizard and her associates confirms the importance of the first two factors specified in hypothesis 17 as essential for counteracting institutional deprivation: that the setting must permit adult-child interaction and provide materials for activities that can be engaged in by the child, alone or jointly with the caretaker. Additional support for this conclusion is found in the results for two of Tizard's comparison groups. The first consisted of previously institutionalized children who had been restored to their mothers by the age of four and a half. All the mothers came from poor socioeconomic backgrounds, and half were single parents. The investigators found that "the mean test results of the children restored to their mother a year previously, at an average age of 3½, were lower, although not significantly so, than those of the institutional children. In exchange for acquiring a mother, they had lost some environmental advantages (e.g., they had many fewer toys and books and were read to less often)" (1974, p. 98).

A different picture was shown by a second comparison group consisting of children from the institutions who, between the ages of two and four, had been adopted, mostly by middle class families. Although there was no evidence of selective placement at the time of adoption, by four and a half years of age these youngsters had a significantly higher IQ than any other group. In the view of the investigators, "these children had acquired not only a mother, but a much richer environment than was provided by the institution" (p. 98).

It would be wrong, however, to infer that the only changes were in the physical aspects of the environment. Tizard and Rees also calculated an index of breadth of environmental experience based on three scores: one for "frequencies of experiences in the adult world," another for "special treats and excursions," and a third for "literary experiences" (p. 96). In general, the four groups showed significant differences in measures of what is perhaps best referred to as the richness of their social environment: "On all our measures of breadth of experience the adopted children scored highest" (p. 98); next in line were children living with natural parents, the majority from poor socioeconomic backgrounds: finally, the environments of institutionalized and "restored" children received the lowest ratings.

The question arises what aspect of the social environment is most relevant to developmental advance. An answer is suggested in the following account of situational changes associated with a cognitive gain exhibited by the institutionalized children between the baseline assessment at age two and the evaluation at age four and a half:

At the age of two we had found these same children to be somewhat retarded in language development; by 4½ years, whether still institutionalized or not, this retardation had disappeared. During the intervening period the institutionalized children had been cared for by a continuously changing roster of staff, but two measured aspects of the environment had improved: the children were spoken to with increasing frequency as they grew older (Tizard et al., 1972), and a more varied range of experiences was offered them. (Pp. 97–98)

Once again there is evidence that, from a developmental perspective, the two most critical aspects of the institutional setting are those features, both physical and social, that enable and encourage the child to participate in a variety of activities both jointly with an adult and spontaneously by himself or with other children. If so, can we conclude that these constitute the necessary conditions to insure the child's normal development, not only in institutions but in other settings as well? This is indeed the conclusion reached by Tizard and Rees:

As far as cognitive development is concerned, institutional life is clearly not inevitably depriving; indeed many of the children must have developed faster than they would have done at home. All the evidence from this and other studies suggests that children who are not often talked or read to and are not given a variety of stimulation tend to be retarded whatever the social setting; institutional retardation, when it occurs, derives from the same poverty of experience as other environmentally produced retardations. (p. 98)

Although some studies do point in the directions indicated by Tizard and Rees, a hard look at the evidence calls for a more qualified statement. Even when restricted, as it properly is, to the sphere of "cognitive development," the formulation is perhaps a bit too firm. The measure of intellectual function employed by Tizard and her colleagues was the Wechsler Intelligence Scale. It is true that the means for all four settings, institutional as well as family-based, were above 100 both on the full and on the verbal scales, and this finding presumably is the basis for the authors' statement that "all the groups were at least average and that there was no evidence of language retardation in the institutional children" (p. 95). Although an analysis of variance of the full scale score on the Wechsler showed a reliable setting effect, this was due primarily to the superior performance of the adopted group. The mean for the institutional group was lower than that for its lower class family control (105 versus 111). Moreover, two years later, when the study was republished in a series of collected papers (Clarke and Clarke, 1976), the original investigators added follow-up data on the children at age eight. Whereas the children from adopted and lower class families had essentially maintained their status (IQs of 115 and 110 respectively), the seven children still remaining in an institution showed a drop from 105 to 98, with all but one member of this group exhibiting the decrease. After ruling out selective attrition as a likely explanation, the authors remind the reader, "Nevertheless, it should be noted that the mean I.Q. of the institutional children is still average" (Tizard and Rees, 1976, p. 148).

But even if one is reassured by this fact, a larger issue arises. In his original monograph Bowlby (1951) concluded, on the basis of the research then available, that when children are exposed to maternal deprivation, "not all aspects of development are equally affected. Least affected is neuromuscular development, including walking, other locomotor activities, and manual dexterity" (p. 20). The most influenced are speech, "the ability to express being more affected than the ability to understand" (p. 20) and "emotional adjustment, in particular the capacity to establish and maintain genuine emotional attachments, these affective disorders not becoming fully apparent until later childhood and adolescence" (pp. 30–36).

The studies examined, particularly those of Goldfarb, and Pringle and Bossio, show evidence clearly consistent with Bowlby's conclusions. Under these circumstances, even Tizard and Rees's restriction of their generalization to "cognitive development" may be too broad. A more precise formulation would have been "cognitive development as measured by standardized intelligence tests." The aspects of development most likely to be affected by early institutionalization lie not in the areas typically measured by standardized psycholgical tests but in the person's behavior in everyday life, particularly in situations requiring initiative and sustained effort and in relationships with other people.

In this regard it is noteworthy that in the most recent publication from the Tizard longitudinal study, the authors (Tizard and Hodges, 1978) report significant differences at age eight in the classroom behavior of previously institutionalized children ($N = 36$) and of those raised exclusively in home settings. The latter consisted of two groups: other pupils in the same class ($N = 36$) and children in the original, noninstitutionalized comparison group ($N = 29$). Although the teachers were not informed of the children's prior background, pupils who had spent some time in institutions were more often characterized as exhibiting antisocial behavior and were described by such terms as "restless," "quarrelsome," "not much liked by other children," "irritable," "attention seeking," "disobedient," "often tells lies," and "resentful or aggressive when corrected." Similar but less pronounced differences were reported by the children's parents at home and by houseparents in the institution.

These results from an English study parallel those found by Goldfarb forty years earlier in the United States. Tizard and Hodges go beyond Goldfarb, however, in relating shifts in child behavior over time to both group and individual differences in the environments in which the children had been living. Thus the highest mean IQ (115) was obtained by previously institutionalized children who had been adopted before the age of four, compared with a mean of 103 for children restored to the natural families at the same age. "This adopted group also had a reading age ten months in advance of that of the restored children" (p. 112). By contrast, institutionalized youngsters adopted or restored after the age of four and a half scored consistently lower in IQ (means of 101 and 93, respectively). Moreover, differences in intellectual performance as well as in social behavior varied directly with the strength of the emotional tie with the child reported by the mother or housemother. The strongest bonds were described by adoptive mothers and those who had raised their own children from birth, the weakest by housemothers in institutions and biological mothers whose children had been restored to them after having been institutionalized. Finally, important from a developmental perspective were significant associations between attachment to the mother, cognitive measures, and "relative lack of behavioral problems" (p. 112). These relations obtained within as well as across social class groups.

On the basis of their findings, Tizard and Hodges conclude that "the subsequent development of the early institutionalized child depends very much on the environment to which he is moved." With respect to aftereffects of institutionalization, the authors take a more qualified stance: "these findings appear to suggest that up to six years after leaving the institution, some children still showed the effects of early institutional rearing" (p. 113). It is important to recognize that the institutions in question are the same ones described earlier by Tizard and Rees (1974) as having "a good staff-child ratio, together with a generous provision of toys, books, and outings ... in the absence of any close and/or continuous relationship with a mother substitute" (p. 97)

In my view the most recent findings of the Tizard group suggest that the absence or disruption of such a relationship is not without some negative developmental consequences. The association the investigators found between indexes of attachment and developmental gains, as well as their finding of developmental disruption among children restored to mothers who had not developed strong attachment toward them, constitutes additional evidence in support of hypotheses I derived earlier. These were based on other research emphasizing the importance of maintaining the continuity of the dyad between the young child and his primary caretaker, and the critical impact of ecological transitions in early childhood (hypotheses 6, 7, and 16 through 18).

Such supportive evidence is by no means conclusive. In their final paragraph, Tizard and Hodges correctly emphasize "that [at eight years of age] the children are still very young and that it is too soon to come to any conclusions about the long-term effects of their early experiences" (p. 117). Nevertheless, the findings of Tizard and her colleagues point to directions for future research. Their work underscores the importance of assessing development over a wide range of human activities—intellectual, emotional, and social—as manifested in the actual settings in which people live. It is also clear from this careful longitudinal study that the issue of the long-range effects of institutionalization, especially in institutions of good quality, cannot be resolved until investigators go beyond the testing room to compare matched samples of previously institutionalized and noninstitutionalized persons as they function in the everyday environments of home, school, workplace and community, for it is here that the distinctive outcomes of differential socialization are most likely to find expression. Once such investigations are carried out, they will, in all probability, show considerable attenuation by adulthood of any effects of institutional deprivation. But short of providing a functional equivalent of a family for each of its residents, the institution, even if it offered children a stimulating and humane environment, is likely to produce some residual deleterious effects in later life.

This qualification is a very critical one. Were Tizard and Rees (1974) to accept it, they would be abandoning the position they have taken which in effect makes the institution and the family equivalent with respect to the possibility of deprivation. In their views the potential exists in either setting for "the same poverty of experience," which can be averted if children are "often talked to, read to, and . . . given a variety of stimulation" (p. 98).

There is indeed evidence that, in some homes, the physical and social environment is so impoverished and chaotic that placement in an institution initiates a period of psychological recovery and growth (Clarke and Clarke, 1954, 1959; Clarke, Clarke, and Reiman, 1958). Nevertheless, the leveling position taken by Tizard and Rees fails to consider the differential properties of homes and institutions as ecological systems. In terms of microsystem elements, the roles, activities, and relations that are typically introduced or encouraged in children's institutions differ appreciably from those that are present or tend to evolve in the home. At the most general level, the institution is a formal structure in which the caretakers are professionals or paraprofessionals, whereas the home is, by comparison, highly informal, and the caretakers are amateurs, whose motives for performing their work are very different. The analysis of research on the nature of interpersonal structures most conducive to human development resulted in a formulation (hypothesis 7) that emphasized "the participation of the person in progressively more complex patterns of reciprocal activity with someone with whom that person has developed a strong and enduring emotional attachment." Elsewhere, I have referred to this requirement as the child's need for "the enduring irrational involvement of one or more adults in care and joint activity with the child" (Bronfenbrenner, 1978b). One of the demands of a professional role is precisely that one must *not* develop irrational involvements: witness the practice in the residential nurseries studied by Tizard and Rees to "discourage ... close personal relationships between staff and children" (1974, p. 97). Moreover, in a family there is only one set of parents caring for children of varying ages, but at an institution changing caretakers work on different shifts typically with children of similar age. As a result, the development of progressively more complex patterns of reciprocal activity in the context of a strong and enduring emotional attachment is not as likely to occur.

Differences between the home and the children's institution are not restricted to the microsystem. At the mesosystem level, the institution is much more isolated from other settings than is the home, so that the child is far less likely to gain experience in other environments. In terms of the exosystem, the personnel and practices of an institution are less susceptible to influence from the external community and less adaptable to modifications and innovations in the interest of the child's transition into other settings. Finally, from the viewpoint of cultural values and expectations, being raised in an institution carries a stigma that can become a self-fulfilling prophecy of failure.

The cumulative effect of these multilevel contrasts can hardly be

negligible. Even though persons subjected to institutional deprivations in early childhood can live normal lives as adults, freedom
from debilitating psychopathology is not synonymous with optimal
psychological functioning. From the perspectives of both science
and social policy, it is necessary to establish the distinctive ecological properties of institutions as they affect the course of behavior
and development. An essential part of such an endeavor should include experimental modifications in the structure of institutions and
experiments in alternative modes of care, that might produce the
ecological conditions most favorable to psychological growth.

It was noted at the beginning of this chapter that studies of the
effects of the institution as a context of human development were
overwhelmingly concentrated on the issue of early deprivation. The
vast majority of the institutionalized population are in fact adults.
They are principally the mentally retarded, the psychiatrically disturbed, the chronically ill, the delinquent, and—especially—the
elderly. Although little research has been conducted on the impact
of institutionalization on these persons, the available evidence suggests some continuity with the findings for children. An experimental program designed by Blenkner, Bloom, and Nielson (1971) to
improve protective services for the elderly was successfully implemented but produced an unhappy boomerang effect. The sample
consisted of 164 noninstitutionalized persons over sixty years of age
who were deemed incapable of adequately caring for themselves
and were being carried on social agency rolls. The cases were randomly allocated to experimental and control groups. The former
were given "considerably greater service of a more varied nature
than was ordinarily available in the community" (p. 489). Four
highly qualified caseworkers were hired to implement the special
program. The control group received the "standard treatment" provided by local social agencies.

Although the experimental group clearly received a higher level
of protective services, findings on measures of functional competence administered after one year failed to show a reliable advantage for the special program. In fact such differences as there were
favored the control group, and mortality rates—referred to by the
authors as "the ultimate deterioration in competence"—showed a
similar trend: 25 percent for program participants versus 18 percent for the controls. This was "a discouraging but not significant
difference. It is clear that the demonstrated treatment did not prevent or retard deterioration" (p. 492). Concerned by the trend, the

investigators continued to gather follow-up data on mortality rates in the two groups. At the end of four years, the survival rate in the experimental group was 37 percent compared with 48 percent for the controls.

Despite the paradoxical character of the findings, the investigators state that they "did not come as a complete surprise" (p. 494). In their view, the key to the paradox lies in the higher rate in institutionalization achieved as a direct outcome of the more intensive casework conducted in the special program. The differential of 34 percent versus 20 percent in the first year persisted after the program was terminated so that "by the fifth year from time of registration more than three fifths (61 percent) of the demonstration participants and fewer than one-half (47 percent) of the control participants had been institutionalized" (p. 495). The higher death and institutionalization rates observed in the experimental sample led the investigators to posit a causal link between the two. They "theorized . . . that because hopelessness could literally kill and because institutionalization generated feelings of hopelessness, whichever group (demonstration or control) had the highest placement rate would also have the higher death rate" (p. 494).

Blenkner and her colleagues then proceeded to test their hypotheses by calculating separate survival rates for institutionalized and noninstitutionalized subjects in both the experimental and control groups. The results were in the expected direction. By the end of the fourth year, the survival rate for institutionalized patients was about 32 percent, with no difference between those who had been in the experimental as opposed to the control group. The rate for noninstitutionalized patients was considerably higher with a marked difference in favor of those who had not been enrolled in the special program (57 percent versus 44 percent). Controlling for age by comparing the observed rates with those expected from standard life tables did not alter the pattern of results. It would appear, therefore, that the major factor accounting for the higher mortality rate in the experimental group was their earlier entry into an institution as the result of the more intensive casework services made available to them on a random basis. Once the subjects were placed in an institution, whether they had previously participated in the experimental program made no difference to their survival. Even short of institutionalization, however, those who received the special program were more likely to die than those who did not. In light of these findings, the investigators conclude on a cautionary note: "These are discouraging facts that should not deter us from further attempts to help. We should, however, question our present prescriptions and strategies of treatment. Is our dosage too strong, our intervention too overwhelming, our takeover too final? Some of the data pertaining to factors predictive of institutionalization or survival suggest we are prone to introduce the greatest changes in lives least able to bear them" (p. 499).

The issue is not merely one of the timing of institutionalization, but of the nature of institutions as social systems, their present isolation from other settings in the society, and the consequences of these structural features for the behavior and development of those who inhabit such ecological niches—either as inmates or as caretakers. For if Zimbardo and his colleagues are correct in asserting that a prison environment affects the guard no less than the prisoner, the same may hold true for attendants in institutions and perhaps for their superiors as well. Social scientists have yet to probe the impact on adult development of the settings in which people spend much of their lives.

The investigations reviewed have served a three-fold purpose. First, they have shed light on the distinctive properties of the institution as an ecological setting that both evokes and inhibits certain kinds of molar activities, roles, and patterns of interpersonal relationships with resulting impact on the course of human development.

Second, these same researches have served to corroborate the general principles of development-in-context embodied in the hypotheses presented in earlier chapters. In particular, Spitz's perceptive exploitation of the existence of children's institutions with contrasting physical arrangements, operating procedures, role repertoires, and organizational structures has provided strong evidence in support of the first seven hypotheses, which assert the developmental importance of engaging the growing child in progressively more complex molar activities, patterns of reciprocal interaction, and what I have called primary dyadic relationships with adults functioning in a parental role.

Furthermore, Spitz's major findings have been supported in a substantial number of independent investigations. These include both comparative studies of natural settings, as exemplified by the work of Goldfarb, Pringle and Bossio, and Tizard and her colleagues, and the planned experiments by Schaffer, Prugh and his associates, and Skeels. This second group of projects, involving changes in role structures, behaviors, and expectations, gives evidence for the power of roles to alter behavior (hypothesis 9), particularly when such roles are given institutional support (hypothesis 10), are complementary to each other (hypothesis 12), and emphasize cooperative activity (hypothesis 13). All these facilitative functions were manifested in the transforming experiments examined, especially in that of Skeels.

Third, the researches reviewed in this chapter provide a basis for formulating additional general principles about determinants and processes of development-in-context. The properties of a setting taken as a whole that are conducive to human development can now be specified.

HYPOTHESIS 19
The developmental potential of a setting is enhanced to the extent that the physical and social environment found in the setting enables and motivates the developing person to engage in progressively more complex molar activities, patterns of reciprocal interaction, and primary dyadic relationships with others in that setting.

In addition, although hypotheses 15 through 18 were derived from research on human development in children's institutions, the conditions that constitute the independent variables in these hypotheses are applicable to many human contexts. For example, the stipulation that an environment for young children offer opportunities for caretaker-child activity, permit locomotion, and contain objects that the child can use in spontaneous activity, (hypothesis 15), pertains as much to a day care setting or a hospital ward as to an institution caring for children on a long-term basis. Similarly, inferences about periods of maximal vulnerability (hypotheses 16 and 18) and prevention and reversibility of psychological damage (hypothesis 17) apply to any environment characterized by reduced stimulation and the absence of persons with whom the child could form a primary dyadic relationship. From this point of view, the hypotheses I have developed regarding children's institutions as contexts of development also constitute tentative general principles for an ecology of human development.

8.

Day Care and Preschool as Contexts of Human Development

Having examined the settings in which most human beings begin their existence, I proceed to examine the first environments that increasing numbers of children in modern industrialized societies enter once they leave home: day care and preschool centers. Perhaps because these settings are more accessible to the world of academe, often being included within it, they have generated a body of research that, while far greater in volume, is from an ecological perspective more limited in substance and theoretical scope than investigations in children's institutions.

The limitations of research on day care and preschool environments are those that derive from what I have called the traditional research model and are manifested in the following features.

1. *The empty setting.* It was already noted that the absence of an ecological orientation in research on human development has resulted in the emergence of a curiously one-sided picture: investigations yield mountains of data about differences in outcomes, or lack thereof but very little information about the settings themselves or the events that take place within them. This imbalance is clearly apparent in the studies reviewed in this chapter. In nine investigations out of ten, the setting is defined primarily by the label attached to the group. Some data may be provided about the caretaker-child ratio at the center and, perhaps, the type of family structure present in the home, but the nature of the differences between the two settings is regarded as self-evident. The features of a setting that have been identified as most consequential for behavior and development—molar activities, interpersonal structures, and roles—are rarely even mentioned.

2. *Ecologically constricted outcome measures.* Despite the volume of outcome data, the variables measured are highly restricted in

164

range. As with most studies of environmental impact on development, research on the effects of day care and preschool settings on children has usually relied on either psychological tests or laboratory measures. For the reasons discussed in chapter 6, these procedures are often of questionable ecological validity in the conditions in which they are employed, particularly when the subjects are infants or youngsters of preschool age; rather than assessing the child's general level of functioning, they may reflect his reaction to a rather specific situation as he perceives it. To be sure, once the investigator understands that situation and its perception, the observed response may become amenable to valid interpretation, but usually within a limited ecological context.

The most serious problem with the range of outcome measures is what is omitted: information about the child's behavior in the situations of everyday life—at home, in the day care center, and on the playground. The unknowns include the activities the child does or does not engage in and the roles and relationships in which he becomes involved with other children, parents, and other adults. These are precisely the domains in which, from an ecological perspective, experience in day care or preschool is most likely to have developmental impact.

3. *Fixation on the child as the experimental subject.* In the traditional research model, the focus of attention is restricted to the experimental subject, who in this case is the child. As a result, few investigators have examined, or even recognized, the possibility that the development of other persons besides children can be affected in important ways by the nature of such care arrangements. Parents are, of course, most likely to be influenced, not only in their child-rearing roles but also in their work, spare time activities, and many other aspects of their lives. Again from an ecological viewpoint I suggest that the impact of day care and preschool on the nation's families and on the society at large may have a more profound consequence than any direct effects for the development of human beings in modern industrialized societies.

Whereas the overwhelming majority of studies on the effects of group settings in early childhood are characterized by the limitations in theoretical and methodological scope I have described, a few investigators have begun to explore previously uncharted terrain. Because the results of their investigations transcend, and sometimes call into question, the now well-established findings of researches using more traditional methods and designs, I shall treat the two sets of studies separately and begin with the latter.

The following summary is based primarily on surveys of the research literature by me and my colleagues (Bronfenbrenner, 1976; Bronfenbrenner, Belsky and Steinberg, 1976; Belsky and Steinberg, 1979). Essentially the same conclusions have been reached by other reviewers (for instance, Ricciuti, 1976).

The conventional studies of day care effects have focused primarily on intellectual outcomes as evaluated by intelligence tests and laboratory measures of cognitive functions (such as memory, concept formation, and problem solving). The following conclusions are based on the results of almost twenty studies comparing matched samples of children with and without experience in center-based or family day care. The validity of the conclusions is considerably strengthened by corroborative findings from the first large-scale survey of home and day care recently conducted for a sample of over three hundred children from poor socioeconomic backgrounds in New York City (Golden et al., 1978).

1. For youngsters from disadvantaged socioeconomic backgrounds, experience in a good quality day care center (licensed or meeting Federal standards) tends to attenuate the declines in test scores frequently observed among preschool children growing up in high-risk environments.

2. No comparable beneficial effect on intellectual performance has been observed among disadvantaged children enrolled in family day care.

3. Among children from average, low-risk socioeconomic circumstances, comparisons of those with and without day care experience have not yielded reliable differences in intellectual performance. It is noteworthy that by far most of the day care centers included in these comparisons were well funded, university-based or connected, and possessed well-trained professional personnel as well as high caretaker-child ratios. Many also included curricula emphasizing "cognitive enrichment." It would appear that for youngsters growing up in families with economic, educational, and social resources, exposure to high-quality day care has little effect on intellectual performance, at least as measured by psychological tests and laboratory procedures.

4. With a single exception (Moore, 1975), there are no follow-up studies beyond the preschool years of children with prior day care experience. Hence it is possible that such differences as have been found may disappear with time or, as Moore suggests, that "sleeper effects" may emerge at later ages, particularly after the child enters school.

The question of long-range effects, left essentially unexplored in research on day care, receives some intriguing answers in studies of preschool intervention. The most unequivocal and revealing findings are those based not on test scores but on measures directly related to experience in the settings of everyday life. The research using traditional indexes presents a less optimistic picture than the one that emerges from employing ecologically-rooted measures.

Even more than in research on day care, studies of preschool intervention have relied on intelligence test scores as outcome measures. I (Bronfenbrenner 1974d) collated and evaluated published test results from a selected group of seven major preschool projects meeting three criteria of research design: availability of matched control groups, comparability of measures, and the availability of follow-up data for at least two years beyond the end of the program.

The short-range effects were consistent with those reported for center day care. In the initial stages of intervention, children from disadvantaged backgrounds showed substantial gains in IQ and other cognitive measures, excelling their matched controls and attaining or even exceeding the average for their age. Two studies with children from middle class homes, however, showed either small gains or none at all. The question arises whether the stronger emphasis on educational activities characteristic of preschool curricula results in more intellectual progress for children with preschool as opposed to ordinary day care experience. No clear answer to this question is possible, since all the preschool studies found had been conducted in well-funded, university-based or university-connected centers with trained staff, high caretaker-child ratios, and programs specially designed to provide cognitive enrichment. Day care projects under similar auspices, while generally dealing with younger children, enjoyed the same advantages and produced IQ gains of comparable magnitude (Lally, 1973, 1974; Ramey and Campbell, 1977; Ramey and Smith, 1976).

But the picture that emerged from the longitudinal data, available only for the selected preschool programs, was rather discouraging. By the first or second year after completion of the program, and sometimes while it was still in operation, the test scores of children in the program began to show a progressive decline, and the gap between experimental and control groups gradually narrowed to a difference of no more than a few points at the last assessment, typically three years after termination. Apparent exceptions to this general trend turned out to be faulted by methodological artifacts (such as self-selection of families in the experimental group). On the basis of these findings, I concluded that the substantial IQ gains initially achieved by group intervention programs "tend to wash out once the program is discontinued" (p. 15).

My conclusion has been challenged by Lazar and his colleagues (1977a) on the basis of their reanalysis of original data on many more children from a larger number of projects, including most of those covered in my review. The measures analyzed were Binet IQs obtained three years after program termination and WISC IQs administered in a special follow-up of children from six projects at average ages ranging from almost ten to sixteen. Of the ten projects for which Binet scores were available, all but two showed statistically significant differences in favor of the experimental group. Most of these differences remained reliable even after adjustment for pretest IQ, measured prior to initiation of the program in all but three of the projects. Significant program effects were also obtained on the full scale of the WISC in the project with the youngest children (averaging nine years, nine months). For the five remaining projects with older children (eleven years, three months to sixteen years, nine months), there were no reliable differences on the full scale (although one group of twelve-year-olds did show a significant effect in performance IQ). The authors interpret these findings as refuting the hypothesis "that early education has only a short-lived effect on I.Q. scores" (p. 61).

Several comments are in order on this score. First, Lazar and his colleagues included in their analyses several projects and comparison groups that I specifically eliminated because of what appeared as critical flaws in research design, resulting in noncomparability between the experimental and control groups in ways that could not be adequately controlled by adjustment for pretest score. But even when these projects are included, the differences in mean Binet IQ as documented by Lazar and his associates (1977b) show a marked decline from slightly over seven points in the first posttest to less than three points three to four years after program completion. Furthermore, in the special follow-up, a significant program effect as measured by the full scale of the WISC was limited to a project involving home intervention exclusively (Levenstein, 1970) and which I had singled out as illustrating the superiority of programs focused on parent-child interaction as opposed to group preschool settings.

Nevertheless, in asserting that "infant and preschool services improve the ability of low income children" (1977a, p. 107), Lazar and his colleagues are, as we shall see, correct, whereas my conclusion, based on earlier data, that the effects of group intervention "tend to wash out" (1974d, p. 155) emerges at best as premature—not because the children were not yet old enough, but because more ecologically valid methods for assessing development-in-context had not yet been applied in studies of preschool intervention. Procedures of this kind had been used with instructive results, however, in research on day care.

The first step in a comparative ecology of human development entails a systematic description and analysis of the settings in which development takes place. It is only recently, however, that researchers have undertaken this task in a methodical fashion. With respect to day care, the first such effort was carried out by Cochran and Gunnarsson (Cochran, 1973, 1974, 1975, 1977; Gunnarsson, 1973, 1978), as the initial phase of a longitudinal study of the development of 120 Swedish children brought up in their own homes ($N = 34$), in family day care ($N = 26$), and in center care ($N = 60$ in twelve different centers). At the beginning of the research, when the children were between twelve and eighteen months old, the groups were carefully matched by sex, age of child, number of siblings, socioeconomic status of the parents, and geographic location of the homes.

Before seeking to assess the effect of the three settings on the children, the investigators conducted observations designed to describe the activities taking place in each location, as well as similarities and differences in the nature of social interactions between children and adults and among children themselves. The following differences were observed in the initial phase of the study, when the children were between one and one and a half years old:

Interactions between adults and child were occurring with considerably greater frequency and duration in the homes and day homes than in the centers, thus providing greater opportunity for socialization by significant adults. The interactions which distinguished homes from the centers were cognitive verbal (reading, labeling, face to face verbalizing) and exploratory in nature. The exploring in the homes involved a child playing with things not designed to be played with (plants, pots and pans, mother's lipstick, etc.) . . .

Where socialization practices differed from the homes and day homes to the centers, the differences involved frequency and focusing of negative sanctioning and restricting by the responsible adults. There were more instances where negative sanctions were applied in the homes than in the centers, and these instances often involved the exploring by the home or day home child of "no, no's" not available to the children in the centers. (Cochran, 1974, p. 4)

In interpreting these results, Cochran rejects the emphasis on the importance of maternal attachment, stressed by Ainsworth and Bell (1970) and others, in favor of an ecological explanation of terms of

the different roles performed by adults in the center and family settings . . . Caregivers are wives and neighbors as well as mothers at home, and the enivornment is organized accordingly. Friends and relatives are received in the home. It may be a display area for parents' prized possessions. Plants and flowers are often within reach. The child has access to the dishwashing detergent, the back stairs and the cat. Opportunities for exploration are more numerous in the homes, therefore, than in the centers, where the single role of the adult is child care and the setting is single purpose in design. (1975, p. 3)

In evaluating his results, Cochran considers the possibility that they may be attributable in part to the greater saliency of the observer in the home with consequently greater impact on both caretaker and child. He argues that, whereas this factor might explain the greater tendency for initiation by adults in the family setting, it does not easily account for increased exploratory behavior by the child in the home environment. Cochran sees the two situations as differentiated primarily in terms of the relative prominence of peers and adults: "The pattern which emerges is one where the child in either type of home setting is being constantly reminded of the saliency of the adult, while the center child's attention is drawn via similar restricting and directing techniques to the importance of appropriate peer relations" (1977, p. 706).

At the same time, Cochran calls attention to many similarities in caretaker-child interaction in the two settings: there were negligible differences in amount of positive or negative reinforcement, degree of affection, or helping behavior. The greatest similarity was between the home and the family day care setting. This fact is in accord with Cochran's general conclusion that the observed differences "are occasioned by variations in setting design, which may in turn be a function of different adult role requirements. Caregivers arrange the near ecology to accommodate single or overlapping role demands (mother/wife/friend) and train children to function appropriately within those physical and social constraints" (p. 707).

Similar findings on the American scene have been reported in an unpublished report by Prescott (1973). The study entailed observations of 112 children in fourteen day care centers ($N = 84$), fourteen family day care settings ($N = 14$), and in their own homes ($N = 14$). The investigator found that "adults in both home-based settings were more available to children than in group care; opportunities for the child to make choices and to control the environment were markedly higher than in group care" (p. 7).

Finally, the New York day care study (Golden et al., 1978), although it did not include any observations in the child's own home, provided comparative data about settings in eleven centers and in twenty family day care programs. The results are summarized as follows: "The Group Day Care programs were superior to Family Day Care programs in the amount of play materials, equipment, and space available to children ... The Family Day Care programs were superior to Group Day Care programs in the caregiver-child ratio, in the amount of social interaction and individual attention children receive from caregivers, and in the degree of positive social emotional stimulation provided by caregivers to children during the noon meal" (p. 148).

How do these consistent differences between homes and day care centers affect the child's behavior in everyday life? First, what happens when a child leaves the home to enter a group setting? This phenomenon relates to the broader issue of the effects of day care on the emotional development of the child. In contrast to the contradictory results of laboratory studies employing the "strange situation" experiment, the findings from a series of observational studies and experiments carried out in natural settings by Schwarz and his colleagues present a coherent picture. In their initial inquiry, Schwarz and Wynn (1971) investigated the factors effecting children's emotional reactions to starting nursery school. They found a difference in the degree of distress exhibited after separation from the mother between children who had ($N = 46$) and had not ($N = 51$) previously spent time in the absence of the mother with a group of other children for at least one hour weekly for one month. Observations were made at the point when the mother left after having brought her child to the nursery school, periodically thereafter throughout the first day, and in follow-up sessions one week and four weeks later. An overall measure of distress at separation (based on such behavior as clinging to the mother, crying, or resisting entry into nursery activities) revealed a significantly higher score for children without prior group experience. No reliable differences in emotional reaction or social behavior were detected, however, less than one hour later, or in the two follow-up sessions at two and four weeks.

The investigators included in their research two important experimental manipulations that were counterbalanced to permit an independent assessment of the effect of each. For a random half of the children, with and without prior group experience, the mothers brought their children to the nursery for a twenty-minute visit with the child's future teacher during the week preceding the start of school. The other half were not given an opportunity for such a warm-up experience. Counterbalanced with this experimental treatment was another manipulation; half the mothers were encouraged to remain at the nursery school for twenty minutes for the first session, while the other half were asked to depart as soon as the children hung up their coats. Contrary to the authors' hypotheses, neither of these strategies designed to reduce distress upon separation from the mother showed significant main effects.

The authors summarize the results and conclusions of the entire study as follows: "children who had had prior group experience on a regular basis outside of the home were less apprehensive about the mother's departure. However, even this difference was not detectable beyond the first 40 minutes of nursery school. These results suggest that most children in *comparable samples* will readily adapt themselves to the nursery school situation without special procedures and that previsits and the presence of the mother are not effective in reducing adverse reactions to nursery school" (p. 879).

In a second study, Schwarz, Krolick, and Strickland (1973) observed a group of twenty three- to four-year-olds who had been in day care for an average of about ten months and had just been transferred to a new center. The twenty controls consisted of youngsters without any prior day care experience matched on age, sex, race, and parents' education and occupation. Observations were made of the children's behavior during the first day at the center, with a follow-up five weeks later. Attention was focused on signs of tension or relaxation, expressions of positive or negative affect, and extent of social interaction with peers. The authors write that

the findings of the present study failed to support the view that the early day care experience leads to emotional insecurity. On the contrary, the Early group exhibited a more positive affective response upon arrival in the new day care setting and tended to remain happier than the matched group of new day care children through the fifth week. If the many hours of separation from home and parents (occasioned by early enrollment in day care) had produced insecurity, one would have expected the Early group to be unhappy, tense, and socially withdrawn or "clingy" in reaction to the uncertainty of being left in a new facility with a lot of unfamiliar adults and children. Instead their initial affective reaction was on the average positive whereas that of the non-day care group was initially negative. The Early group, rather than being withdrawn or "clingy," exhibited a high level of peer interaction, significantly higher than that of the Late group and tended to be less tense than the Late group . . . It may be concluded that no evidence was found for the proposition that infant care with its attendant separation from the mother leads to emotional insecurity. On the contrary, early-day care subjects were more comfortable upon entering a new group care setting than non-day care subjects. The greater security of the Early group may have derived, in part, from the presence of peers to whom they had developed strong attachments. (Pp. 344–346.)

A closer examination of the procedures and data in these studies leads to an important clarification. In the first study, significant effects as a function of prior group experience were short-lived and no longer detectable even after the first forty minutes of nursery school. Moreover, in the second study of adjustment to a new day care environment, significant differences in emotional reaction were found only on the first day while the newcomers were hanging up their coats and for two minutes thereafter. It was only during these initial few moments that the home-reared children clearly "cried," "pouted," "whimpered," or "expressed dislike" more than the matched controls with day care experience. Thereafter, the between-group differences in measures of distress were much smaller, approaching significance only on ratings of tension. It would appear that, at least in terms of anxiety level, the children without prior day care experience were adapting to the new environment. Differences in the fifth week were still reliably found in social interaction, defined as engaging in actions eliciting a response from others. It would appear, therefore, that early entry into day care (from about ten months of age) had its major impact not so much in the emotional as in the social realm.

Support for this conclusion comes from the third in the series of studies by Schwarz and his colleagues (1974). As described above, the data consisted of ratings of nineteen matched pairs of three- to 4-year-olds on nine behavior scales four months after the children had been enrolled in a new day care center and again four months later. It was the first substitute care experienced by the home-reared youngsters, while the others had been in group care at another center since about nine months of age.

The two groups differed significantly on three of the nine scales. Most markedly, day care children exceeded their home-reared counterparts in measures of aggression, both physical and verbal, toward both peers and adults. They were also less cooperative in relation to adults and engaged in more running about than sitting in one place. An additional difference, significant at the 10 percent level only, suggested greater tolerance for frustration among home-reared children (as reflected in the ability to accept failure and to be interrupted). While this investigation by Schwarz and his colleagues detected no difference between the groups in the ability to get along with peers, another observational study of what appear to be the same children at about the same time (Lay and Meyer, 1973) indicated that the day care youngsters interacted more with age-mates than with adults while the opposite was true for the children brought up at home. There were also some indications that the children with prior day care experience (who had all been previously enrolled in the same center) exhibited more positive social interactions and tended to socialize more with their own group. Finally, Lay and Meyer found that, compared with home-reared children, three- to four-year-olds who had been in all-day group care for most of their lives spent more time in the large-muscle activity area of the center and less time in the expressive and cognitive areas.

Schwarz's conclusions have been challenged by Macrae and Herbert-Jackson (1976), who replicated Schwarz's investigation using similar scales but obtaining results opposite to his: early enterers got along significantly better with peers and showed a nonreliable trend toward greater cooperation with adults. Macrae and Herbert-Jackson fail to give due weight, however, to the fact that children in their study were appreciably younger (two year old rather than three to four), and that early entrants had been enrolled for a much shorter period of time (thirteen months as opposed to two and a half years). Their sample was also considerably smaller (eight pairs as opposed to nineteen) and, unlike Schwarz's, had not been pair-matched on parental occupation and education.

A considerable body of evidence consistent with Schwarz's findings has been accumulating from a variety of sources. McCutcheon and Calhoun (1976) report that the increased interaction with peers observed in their day care sample was accompanied by a decrease in interaction with adults. In line with this finding, Prescott (1973) found in an observational study that instances of aggression, rejection, frustration, and experiencing pain occurred significantly more often among children in all-day group settings than among those in full-time family day care or half-time nursery-home combinations.

In a similar vein, Lippman and Grote (1974), in a matched sample of 198 four-year-olds cared for in licensed day care centers, licensed family day care homes, and their own families, assessed cooperative behavior in two games in which children from similar day care arrangements were paired as partners. In the first game, requiring spontaneous help to open a box with four spring latches, there were no significant differences by type of care. In the second, involving a choice of a cooperative or a competitive strategy in playing marbles, the home-reared children were more likely to use the winning strategy of taking turns.

In an observational study with a somewhat older sample of middle and upper class first graders in New York City, Raph and his colleagues (1968) found that negative interactions with teachers (but not with peers) varied directly with the amount of prior exposure to group experience (from one to three years) in nursery and kindergarten.

Raph's finding raises the issue of long-term effects. To date, the only research to have followed effects of substitute care beyond the preschool years is a longitudinal investigation conducted in London by Moore (1964, 1972, 1975). The investigator compared development in two groups of children up through fifteen years of age. The first group consisted of forty-eight children who had been cared for during most of the day by someone other than the mother for at least one year before the age of five. The substitute care could have been received either at a center or in a home, so these two features were confounded. The care began at an average of three years of age and continued for a mean of twenty-five months. The home care group consisted of fifty-seven children who had been under the full time care of their mothers, apart from occasional baby-sitting, until the age of five. Both maternal employment and nursery school attendance were excluded. Quite appropriately, Moore refers to this group as having received "exclusive mothering."

Eliminated from both groups were children from single-parent families, as well as those in substitute care for less than twenty-five hours a week or less than one year. All the research subjects had been selected from a larger sample in such a way as to match groups as closely as possible with respect to the following characteristics: sex, age, birth order, and IQ of the child, mother's education and age at child's birth, and father's occupation. The final result of this matching process was quite satisfactory.

Two major types of outcome measures were obtained: mother's responses to a seventy-item inventory of her child's everyday behavior, administered when the child was six, seven, nine, eleven, and fifteen years of age, and ratings by the psychologist of the child's performance in a series of standardized play situations, made on twenty- to thirty-item scales at four intervals between the ages of six and fifteen. In addition, evaluations were available of the child's reading proficiency at age seven and of attendance at school, performance on final examinations, and expressed interests at age seventeen.

In the first assessment (1964), when the children were six years old, those who had experienced substitute care were judged significantly more self-assertive, with both other children and their parents, less conforming and less impressed by punishment, less averse to dirt, and more prone to toilet lapses than their home-reared counterparts. The differences by mode of care, however, were far more pronounced for boys than for girls. Moreover, the contrasting patterns increased in magnitude as the children grew older, with progressively more reliable differences appearing at later ages. Moore refers to this phenomenon as a "sleeper effect" (1975, p. 257). In the latest report (1975) documenting the status of the children at ages fifteen and beyond, Moore found it necessary to summarize separately the effects of early care experience in the two sexes.

Compared with sons raised primarily in their own families, teenage boys with a history of substantial substitute care were more likely to be described by the mother on a behavior checklist as telling lies to get out of trouble, differing with parents about choice of friends, using parents' possessions without permission, and taking "things they knew they should not have" (p. 258). After reviewing all the available information on this group, the author presents the following composite picture: "The . . . boys as a group seem well described by the label . . . 'fearless, aggressive nonconformity.' This involves outgoing interaction with peers on the one hand and numerous differences with parents on the other. The independent ratings of the two psychologists . . . confirm the active and aggressive quality of these boys' behavior" (p. 257).

Boys who had been raised primarily in their own families were described on the checklist by items like the following: "Can be trusted not to do things they should not do," "slow to mix" with other children. Compared with their counterparts with a history of substitute care, the boys themselves expressed a stronger interest in academic subjects, "making or repairing things," and "creative skills" (p. 258). The boys also read significantly better at age seven (when a reading test was given), and were more likely to be still in school and pass their final examinations at age seventeen.

The differences among girls were in a similar direction but much less marked. The girls experiencing substitute care before the age of five "revealed more aggression and ambivalent feelings." They showed more confidence in the standardized task situations, "but in contrast to the corresponding group of boys, their adolescent interests are domestic, not adventurous; while looking forward eagerly toward marriage, they express worries about being seen undressed, and sometimes about leaving home, even occasional nostalgia for childhood, again like the opposite group of boys. But many of these tendencies are of only borderline significance" (p. 260). The girls raised exclusively at home showed something of an opposite pattern. While indicating domestic interests at age eight, by adolescence they were seen by others as more concerned about being popular, described themselves as being active, and expressed a positive attitude toward sex.

In interpreting these results, Moore thought they might be attributable to differences in the personalities of mothers who did or did not place their children in substitute care. The mothers' responses to interviews and questionnaires did indeed reveal some differences between the two groups, which again varied systematically by sex of child. In interviews conducted while the children were still of preschool age, mothers of boys raised at home "were consistently assessed as more anxious and ego-involved with their children, and the level of anxiety increased through the preschool period. This was not true of the mothers of girls." In a questionnaire on parental attitudes (PARI) administered when the children were eight years old, these same mothers gave responses reflecting "a coercive attitude toward boys" but "a tendency toward seclusiveness" with daughters (p. 261).

In an attempt to assess the relative importance for the child's development of maternal attitudes and early care arrangements, Moore carried out analyses of variance using as dependent variables those factor scores of child behavior that had shown significant differences by mode of care. "When the variance was broken down into that due to the regime and that due to associated characteristics of maternal personality, the regime . . . was found to be the primary —and only statistically significant—factor" (p. 262). Moore also carried out an additional analysis to determine whether group differences in behavior might not already have existed when the children were still infants rather than having been produced by their subsequent experiences in child care. The results were negative: it appears that the differences began to emerge at about age three, just when "nursery school began for many" (p. 264).

Moore draws conclusions that, in effect, invoke a plague on both houses and call for a middle ground:

1) Where a mother keeps her child in her own care full time to the age of five (to the exclusion of nursery school as well as of other substitute care) the child tends early to internalize adult standards of behaviour, notably self-control and intellectual achievement, relative to other children of equivalent intelligence and social class.

2.) For boys, this effect tends to persist into adolescence and involves anxiety for adult approval, with consequent inhibition of assertive behaviour, fear of physical hurt and timidity with peers.

3) Mothers who adopt this policy with boys tend to become anxiously ego-involved with them and to profess belief in a coercive rearing policy, but the effects are associated with the restrictive regime independently of any such tendencies in the mother.

4. Where mothering is diffused by substitute care of any kind for most of the day starting before the fourth birthday, boys come to care less for the approval of adults and more for that of their peers; their behaviour tends to become active, aggressive, independent and relatively free from fear despite some adolescent worries, and they are less likely to stay on at school and study for examinations.

5) In girls, the effect of regime as such is less clear. Exclusive mothering seems to involve for them less anxious inhibition than for boys . . . In adolescence it is the exclusively mothered group that appears more outwardly, and the diffusely mothered group more domestically oriented.

6) There are indications that for girls exclusive care facilitates identification and modelling in accordance with the mother's personality . . .

7) Instability of regime introduces cumulative stresses that are likely to be detrimental to personality development.

8) There is some evidence from other research that a compromise regime of stable part-time substitute care from or after the third year may produce the best personality balance, but the best solution for a particular child and family will involve consideration of many individual factors.

9) The effects of group vs. individual substitute care at various ages and of its combination with varying amounts of contact with the mother (and the father) need further investigation, taking account of the nature of the interaction between the child and others under each available regime.

10) For research purposes, outcome may best be measured, not in

terms of good or bad adjustment, but of specific directions of personality development, any of which may be valuable in moderation but detrimental in extreme form. Society needs different kinds of people: it would seem that they can be to some extent produced by varying the mothering regime. But whereas there is no single optimal personality type, there are limits in both directions beyond which deviations become maladaptive. Those limits, and the conditions leading to their infringement, have still to be defined. (P. 270)

It is regrettable that Moore was unable to include the very "compromise regime" that he advocates, since it proved impossible to find a group of children who had experienced stable part-time care for less than twenty-five hours per week, because of the "transitory nature of most such arrangements in the district at the time when the data were collected."

The fact remains, however, that most of the children included in the day care samples of the other studies reviewed here were not receiving the kind of "compromise regime" that Moore advocates of part time care preferably deferred until the child is at least three years old. As I have documented elsewhere (Bronfenbrenner, 1975, 1978b), the national trend is in exactly the opposite direction and is accelerating rapidly. Serious consideration should therefore be given to a general conclusion reached by one of the principal researchers in this field. Having reviewed their own and others' findings, including Moore's longitudinal data, Schwarz and his colleagues arrive at the tentative judgment that "early day care experience may not adversely affect adjustment to peers but may slow acquisition of some adult cultural values" (1974, p. 502).

The allusion to cultural values calls attention to a more general phenomenon that is especially pronounced at least in contemporary American society. Over the past two decades, I along with colleagues both at home and abroad have conducted a series of comparative field studies and experiments concerning socialization by adults as opposed to peers in the United States, the Soviet Union, Britain, Israel, and other industrialized societies (Bronfenbrenner, 1961, 1967, 1970a, 1970b; Devereux, Bronfenbrenner, and Rodgers, 1969; Devereux, Bronfenbrenner, and Suci, 1962; Devereux et al., 1974; Garbarino and Bronfenbrenner, 1976; Kav-Venaki et al., 1976; Lüscher, 1971; Rodgers, 1971; Rodgers, Bronfenbrenner, and Devereux, 1968; Shouval et al., 1975). These investigations, as well as indirect evidence from the experiments of Milgram, Sherif, and Zimbardo discussed earlier, indicate that, depending on the goals and methods involved, experience in group settings can lead to consequences ranging from delinquency and violence through responsible cooperation to unquestioning conformity. Moreover, the cross-cultural studies suggest that peer groups in the United States, while far from either extreme of behavior, are closer to the violent end of the continuum. The tendency of the peer group in a given culture to predispose children, especially boys, toward greater aggressiveness, impulsivity, and egocentrism appears to be associated with an ideology of individualism and a social structure that emphasizes segregation by age. Both these features are prominent in contemporary American society and, to a lesser extent, in Britain as well. This may explain the continuity between the findings of American studies and Moore's research in London.

If this interpretation is correct, it would imply that the somewhat higher aggressiveness observed in children with day care experience in the researches we have examined may not be a consequence of early upbringing in group settings per se but rather a reflection of the general role of peer groups as contexts of socialization in some cultures. For day care can—and does—have quite different developmental consequences in Soviet society, leading children not toward aggressive individualism but rather to conformity and compliance (Bronfenbrenner, 1970a, 1970b). Still another variant occurs in Israel, especially in the collective settings of the kibbutz and moshav, where the outcome of group care is a blend of independence and cooperation (Avgar, Bronfenbrenner, and Henderson, 1977; Kav-Venaki et al., 1976; Shapira and Madsden, 1969).

It is interesting in this connection that the only comparative study of children raised in day care as opposed to at home that failed to find setting differences in antisocial behavior is Cochran and Gunnarsson's longitudinal study in Sweden (Cochran 1977; Gunnarsson, 1978). In a follow-up when the children were five and a half years old, observations were conducted either in the center or in the home depending on which had been the primary context of upbringing since the first year of life. As in Moore's research, the results revealed a reliable setting effect only for males: "Boys observed in the center interacted less with adults and much more with peers, while with girls the interaction patterns were quite similar in the two settings" (Gunnarsson, p. 68). The difference in level of interaction was not expressed in greater negative affect or aggressiveness, but rather in such behaviors as using the adult as a resource, which was significantly lower for boys in day care than for their counterparts raised at home. In his report on their joint study, Gunnarsson emphasizes the contract between his and

Cochran's results and those reported by other researchers (including those cited earlier in this chapter): "Our observational data coupled with our clinical experiences, based on hundreds of visits in day care centers and home settings, do not support these previous findings" (p. 103).

Gunnarsson might have gone on to say that his results actually contradicted previous findings. One of the reliable setting differences found in the Swedish study was in the frequency of observed cooperative behaviors at age five and a half. *"Center boys were more likely than boys in the homes, and girls in either setting, to be engaged in cooperative activities and information sharing. They were also found complying with peers more than the home boys did"* (p. 97).

The finding suggests that, whereas day care experience tends everywhere to enhance conformity to peer norms, the content of these norms is a function of more general cultural values. In their conclusion of a comprehensive review of day care research, Belsky and Steinberg write, "Like all social and educational efforts, day care programs are likely to reflect, and in some measure achieve, the values held explicitly or implicitly by their sponsors, and, through them, by the community at large" (1978, p. 942).

There is one respect in which Gunnarsson's results in Sweden are paralleled in many of the other countries studied. One of the most pervasive findings in the follow-up of the children at age five was the contrasting behavior of boys and girls. Sex differences were "more pronounced than differences found between different child care environments" (p. 2). Moreover, as in Moore's longitudinal study, the setting effects that did occur were markedly differentiated by sex of child, with boys being affected more than girls. Although few interactions by sex have been found in day care research with children under five, such effects are very frequent in the cross-cultural experiments on the susceptibility of children to influence by peers versus adults. While the direction of main effects by source of influence sometimes differed from culture to culture, interactions by sex showed a consistent pattern, with boys being more affected by environmental contrasts than girls.

Although this is a post-hoc finding that requires cross-validation in other domains, it is sufficiently pervasive to demand explanation. One explanation is suggested by Gunnarsson's speculation regarding the cause of the sex difference emerging in his own study. Citing a recommendation of the Swedish Child Care Commission urging recruitment of male staff members in day care centers, Gunnarsson comments: "Our data show, indeed, that with no men in the day care centers (and, of course not, in the homes!) homes and centers do not differ much in the way that they assist in conserving old sex-role stereotypes" (p. 100). It is possible that the availability of a same-sex role model for girls in both home and day care settings enables them to maintain more stability in moving from one setting to the next. Compared with the highly controversial notion of biologically based sex differences in susceptibility to environmental effects, a hypothesis founded on role models is more amenable to empirical and, indeed, experimental test.

Having reviewed the evidence on the effects of day care in a variety of domains, we are in a position to evaluate what aspects of behavior and development in what life settings are most likely to be affected by the day care experience. We observe first that the most salient differential effects of mode of care have been recorded when children brought up in different settings are observed in the *same* setting—in the day care center or, especially, in a home. It may be significant that the only two studies failing to find substantial setting differences in aggressive and antisocial behavior as a function of mode of care employed comparisons in which the children were observed in settings of quite different types: the home-reared or family day care group was observed only in the home, whereas the children experiencing group care were watched only in the center. This procedure was followed in the New York City survey (Golden et al., 1978) as well as in Cochran and Gunnarsson's longitudinal research. Indeed, in evaluating the Swedish investigation at the conclusion of his follow-up report, Gunnarsson identifies this difference in locale of observation as the principal shortcoming of the research: "In our view, the major limitation of the study is *the absence of observations in the homes of the center children.* Data on social interaction patterns in the home environments would have contributed to a fuller understanding of the child's social experiences" (p. 105).

From an ecological perspective, Gunnarssons' statement on the importance of home observations in research on effects of day care touches on a vital methodological, substantive, and theoretical issue but does not go far enough. In terms of research design, observing each group only in its own setting fails to meet the criterion of developmental validity (definition 9). The differences observed may simply represent adaptation to a particular situation and reflect no lasting influence, since the behavior of the two groups might become exactly the same once they are placed in the same setting.

The foregoing methodological issue leads directly to a substantive concern.

PROPOSITION H
If different settings have different developmental effects, then these effects should reflect the major ecological differences between the settings, as revealed by contrasting patterns of activities, roles, and relations.

Although in their investigation Cochran and Gunnarsson failed to follow the basic strategy employed by most investigators of observing children with and without day care experience in the same setting, they did something equally essential that no researchers in day care had done before them: they performed a comparative analysis of the two kinds of settings. In our terminology, the settings differed primarily in two elements of the microsystem: molar activities and relations. Whereas similar or at least analagous roles existed for both children and adults, these participants did rather different kinds of things in the two settings and tended to form different kinds of dyads. As has been said, in the homes there was more reading, labeling and face-to-face interaction, as well as more exploratory behavior. In the interpersonal sphere, adult-child dyads were much more frequent in the home, whereas interactions between peers predominated at the center; there was also a corresponding difference in balance of power, with more exercise of authority, especially in terms of restrictions, in the home than at the center.

Data from the New York City study indicate that setting differences in the United States are similar to those in Sweden, with one possible exception. Information about homes was limited to family day care settings. These differed from centers in having less space, play materials, and equipment available for the children but a higher caretaker-child ratio and more individual attention given to children by the staff. A possible departure from the Swedish pattern is the prominence of "positive emotional stimulation" provided by the caregivers (Golden et al., p. 148).

Given these distinctive setting properties, our thesis would lead to the prediction that day care experience would show its most powerful and enduring effects in the future content of molar activities engaged in by the child and in the changed character of the child's relations with adults and peers. These are of course precisely the areas of greatest contrast found in Moore's follow-up study of the behavior at ages six through seventeen of children, especially boys, who experienced differential modes of care before the age of five. To be sure, most of the significant differences reported by Moore were based not on direct observations but on the mothers' responses to an inventory of the children's behavior, presumably mostly at home. It is therefore reassuring that setting differences in patterns of interaction with peers and adults primarily for boys were also found in the follow-up at age five in the Swedish study. Unfortunately the observational data collected did not include the content of molar activities engaged in by the children. In retrospect, Gunnarsson, with characteristic perspicacity, reports this as his next most serious error of omission: "A second limitation ... has to do with the absence of an activity sector. We have collected data on the social experiences of our children, but we lack systematic descriptions of the kind of things our children were doing, whether interacting with adults and peers or not" (p. 106).

In acknowledging the need for home observations on children in center care, Gunnarsson emphasized the potential contribution of such data "to a fuller understanding of the child's *social* experiences" (italics supplied). But the difference in the pattern of molar activities engaged in by children in homes versus centers as described by Cochran (1974) is focused around "cognitive verbal" and "exploratory" activities including "reading, labeling, face-to-face verbalization" and "playing with things not designed to be played with (plants, pots and pans, mother's lipstick, etc." (p. 4).

There is another consideration, not mentioned by Gunnarsson, that argues for the importance of home observations on children receiving center care: our theoretical model leads to the prediction that this experience will affect not only the behavior and development of the child but also that of the parents, especially the mother. A number of investigations report findings consistent with this hypothesis. Lally (1973) reported that the number of high school diplomas earned by mothers whose children were cared for in a university-sponsored day care center was significantly greater than for their counterparts raising children at home. Other investigators have found that as satisfaction with substitute child care increased, so did marital satisfaction (Meyers, 1973) as well as the working mother's attitude toward the job (Harrell, 1973; Harrell and Ridley, 1975). Unfortunately, since none of these studies employed a before-and-after design, a strong possibility remains that the observed differences were not a function of the type of care arrangements but of other, contemporaneous social and economic changes.

Evidence for the occurrence of such changes is found in the Swedish longitudinal study. At the outset of their research, Cochran and Gunnarsson had carefully matched groups of children receiving different forms of care by sex and age of child, number of siblings, socioeconomic status of the parents, and geographic location of the homes. "Even parental attitude toward the day care center and toward the maternal role was taken into consideration through selection of 'home families' from center waiting lists" (Gunnarsson, 1978, p. 90). But four years later, at the time of the follow-up study, Gunnarsson found that the families were no longer comparable on most of these factors. Moreover, the changes were not random, but varied systematically as a function of mode of care.

While different factors may have played a predominant role in the life of each individual family, our groups differed systematically on four variables. *Children in center care were more likely than home children to have only one parent in the home.* They were also *more likely not to have brothers and sisters*, and *more likely to live in apartments* than the home children who often had moved with their parents to suburban private homes. Surprisingly enough, we also found more girls than boys in the center sample and substantially more boys than girls in the home group at phase two, a finding which argued for the importance of including "sex of child" as a separate variable in the data analyses. (P. 91)

To investigate possible reasons for these changes, Gunnarsson made inquiries of the parents. Several explanations were offered. One was that parents who wished to enter their children in day care were not able to do so because of administrative obstacles (such as lack of places, preference given to single mothers, requirement of full time maternal employment). A second was that having more than one child complicates the problem of finding arrangements for substitute care because of lack of places. The mother may also decide to stop working and stay at home to care for all her children. A third explanation was that moving from an apartment to one's own home correlates both with family size and with the family's financial resources. As a result, "single-parent families and two-parent families with only one child have tended to remain in relatively less expensive and smaller, apartments." (p. 92).

In Gunnarsson's view, these findings carry an important lesson for research design.

These detected differences should be considered important contributions to the understanding of human development, rather than being looked upon as "scientifically inconvenient." It is the strength of the longitudinal design that those "changes over time" that are bound to occur are recognized and taken into consideration when outcome data are analyzed. All too often has research through "one-shot designs" failed to shed light on the actual circumstances prior to and after "the experiment," hence leaving us with the almost impossible task of trying to figure out the truth *behind* available test-scores or context-free performances. Implicit in this line of reasoning is the notion that one has to look with suspicion on those longitudinal studies where the participating families (children, neighborhoods) over long periods of time continue to be neatly comparable on a single variable at study. Nature did not create the word "confounding variable," Man did. (Pp. 92–93)

While there is wisdom in Gunnarsson's words, they should not be interpreted to imply that change is inevitable. His analysis reveals a profoundly important ecological principle: persons who find themselves, for whatever reason, in given environmental settings are thereby often set on certain life trajectories, not because of the internal properties of the settings themselves but because of the position of the settings in the larger context of meso-, exo-, and microsystems. This inexorable principle is tragically illustrated in Furstenberg's longitudinal study of unwed teen-age mothers (1976). Once such a girl becomes pregnant, much of the rest of her life is foreordained and, indeed, foreclosed in terms of future education, work opportunities, income, marriage, and family life.

But the life course may also remain on an even keel, as shown by findings from the New York City survey of infant care (Golden et al., 1978). Contrary to the results obtained in American studies cited earlier, as well as in the Swedish longitudinal project, this large-scale investigation involving over 300 children in eleven centers and twenty family day care programs, as well as a comparison group raised in their own homes for the first two and a half to three years of life, found no differences over time among groups, either in the families' life circumstances as reflected in socioeconomic status, income, and family structure or in mode of family functioning as measured on an instrument designed by Geismar and Ayres (1960) to diagnose status and improvement of families in response to casework intervention. "While there were increases in some of the measures over a period of several years, they were not related to the type of day care programs or length of time in the program" (Golden et al., 1978, p. 156).

Two comments are in order regarding this striking lack of relation. First, possible group differences may have been attenuated by the way in which the home-reared sample had been selected, since it consisted of children who, though raised at home until the age of at least two and a half, were entered into one or another form of day care thereafter. A second explanation emerges from a more detailed analysis carried out by Golden and his colleagues of the backgrounds and psychological characteristics of the families in their sample. They summarize their findings as follows:

It is our impression that public infant day care services in New York City are used by relatively intact, fairly well functioning, poor working families. They work even though they may not earn much more than they would get from public assistance. They work whether public infant day care services are available for their children or not. This picture of poor working Black and Hispanic families runs counter to the stereotypes one often reads in the literature about disorganized, minority welfare families. (Pp. 157–158.)

Larger ecological systems can manifest stability as well as change. When they do so, they can lend stability to the settings they contain and to the human beings who live within them, even, as in this instance, when social, economic, and ethnic stereotypes presume the contrary.

It would be highly desirable to be able to cross-validate our tentative conclusions about effects of day care against analagous results from research on preschools, with due regard for systematic differences between the two types of settings. This analysis is possible only to a limited degree, principally because studies on preschools as developmental contexts are much more restricted from an ecological perspective—in research design, comparative analysis of settings, and outcome measures. Thus it has not been possible to locate any study of the effects of preschool experience on behavior in the home. Nor is there a preschool counterpart of Cochran's comparative analysis of home and day care. Outcome measures are narrowly restricted to intelligence and achievement tests. There is no research evidence bearing on the question whether early entry into preschool has effects similar to center care in predisposing children toward aggression, egocentrism, and antisocial behavior. Given the breadth and variety of the preschool curriculum compared with that of the elementary grades, one might expect preschool attendance to increase the range of molar activities engaged in by the child at home and in other settings outside the preschool center; this possibility also remains unexplored.

There has been a small yet significant breakthrough in the use of more ecologically valid outcome measures. In their long-term follow-up of fourteen experiments in preschool intervention, Lazar and his colleagues (1977a, 1978) have managed ingeniously to find real-life events that could serve as indexes of the long-term effectiveness of intervention programs conducted when the children were still of preschool age. The measures selected were "two indicators of actual school performance . . . whether a child was held back in grade and assignment to special education classes." In the authors' view, such measures have "one major advantage over the use of I.Q. and achievement tests in that grade failure and special education placement are concrete indicators of whether a child has performed acceptably within his/her educational institution" (p. 62).

To assess program effects on grade retention, Lazar and his colleagues examined the school records of the children who had participated in seven major preschool intervention projects involving 544 program children and 246 controls. At the time of the follow-up, the children's placement ranged from the third to the twelfth grade, with most being in the seventh and eighth grades. Since the number of cases in the program and control groups varied markedly from project to project, the authors calculated percentage figures separately for each and used an appropriately weighted statistical procedure for calculating significance levels. The analysis revealed a reliable program effect. A rough indication of its magnitude is provided by the overall percentages of children held back in each of the two groups—17 percent for those who had been enrolled in preschool intervention as opposed to 24 percent for the controls.

The results for assignment to special education were somewhat more pronounced. Relevant data were available for children from five projects with 320 enrolled in programs and 141 serving as controls. Again there was a significant program effect, the figures being 13 percent for those receiving preschool intervention and 28 percent for the controls. These differences remained significant after control for IQ obtained before the program began (Vopava and Royce, 1978).

On the basis of these findings, Lazar and his colleagues conclude that:

the combined results from all projects indicate that early education helps low-income children to meet the minimal requirements of their schools . . . Thus it appears that early education can result in cost savings by reducing the rate of assignment to special education and/or the rate of grade failure. More importantly, there is now evidence that early education can improve the probability that low-income children will be able to perform acceptably in school and not become labeled as failures. (P. 73)

Even though the percentage differences are substantial for only one of the two outcome measures—placement in special programs —Lazar's findings are indeed significant both in scientific and in human terms. The discovery that exposure to an enriched environment in the preschool period can set in motion forces that persist into succeeding years extends our understanding of the resilience of the young human organism and of the momentum of developmental processes once they are set in motion. At the level of the individual and the family, whether a child is able to progress normally in school may determine his subsequent life course. For these reasons, it is especially important that analyses of the kind that Lazar and his colleagues have undertaken be carried out on as firm a scientific footing as possible.

Recognizing that one of the most troublesome problems in longitudinal research is posed by loss of cases over time, Lazar and his associates carried out a second analysis: they examined rates of attrition separately for the experimental and control groups in each project. Although the proportions for the two groups turned out to be similar, the losses in the follow-up study were appreciable. With respect to the School Record Form, for instance, from which the information on retention and assignment to special classes was taken, percentages of cases lost for particular projects used in the analysis ran as high as 71 percent, with a median figure of 31 percent (Lazar and Darlington, 1978). The next step was to determine whether attrition had operated to introduce biases in the residual samples and in the comparability of their experimental and control groups. Attrition rates proved to be uncorrelated with program effectiveness. For three key background variables—socioeconomic status,, mother's education, and pretest IQ—a two-way analysis of variance design was employed to detect any differences between dropouts and surviving cases, experimental and control cases, and the interaction between the two (differential attrition). Few of these differences were statistically significant, and inspection of the means revealed no consistent trends. Nevertheless, Lazar and his colleagues reran their analysis of program effects, controlling first for pretest IQ and them simultaneously for a whole array of demographic variables including mothers' education, family size, and type of family structure. The differences between experimental and control groups were still significant (Lazar and Darlington,

1978). Thus there appears to have been little bias, at least with respect to the variables for which background data were available.

Still the obtained results must be interpreted with some caution. Even though the overall trends were statistically significant, the several programs were by no means uniformly successful with respect to the two outcome measures. Whereas four out of five projects had shown reliable effects ($p \leq .10$) in terms of assignment to special programs, only one out of seven did so for percentage of grade failures, with one sample even showing a nonsignificant reversal in favor of the control group. The reason for such variation remains unknown.[7] In view of their importance to public policy, the welcome findings of this important study should be regarded as tentative until they are replicated in other experimental programs using additional outcome measures that are as consequential and ecologically valid as those devised by Lazar and his colleagues.

One important question remains to be explored: what specific aspects of a day care or preschool program enhance or impair its effectiveness? Principally on the basis of conventional wisdom, the criterion that has been most widely applied for evaluating adequacy is the caretaker-child ratio. It is only recently, however, that any systematic studies have been conducted on the effects of this variable on the behavior and development of children in group settings. The paucity of research on this factor is all the more curious since it is so readily susceptible to experimental manipulation. To my knowledge, only one such experiment has been conducted.

The most definitive findings on the developmental effects of caretaker-child ratio come from a large-scale project, the "National Day Care Study," conducted by Abt Associates (Travers and Ruopp, 1978) under contract to the Administration of Children, Youth, and Families. The major objective was "to determine the impact of variations in staff/child ratio, number of caregivers, group size and staff qualifications on both the development of preschool children and the costs of center care" (p. 1).

In terms of magnitude, the investigation, which is still in progress, is impressive: "As of January 1978, the study's staff have observed and tested 1800 children, interviewed 1100 parents, observed and interviewed caregivers in 129 classroom groups, and gathered program and cost data from 57 centers located in Atlanta, Detroit and Seattle (sites selected to represent both geographic and center diversity)" (p. 1).

The work is also noteworthy for the breadth and ecological validity of the measures used. Program characteristics assessed included not only easily obtained information on staff-child ratio, group size, and caregiver qualifications but also systematic descriptions of the physical environment and observations of caregiver behavior both in terms of content of activity and the number of children toward whom the activity was directed.

Most impressive of all are the extensive observations of children in the center setting.

Observers coded child behavior in three areas: the degree to which the child was involved in group activities and the nature of those activities; the degree to which the child initiated interchange with other children and how she/he did so; the degree to which the child received input from others, the nature of the input and the child's reaction to it. Examples of the 54 behavior codes included in the instrument are: "considers, contemplates;" "offers to help or share;" "cries;" "asks for comfort;" "refuses to comply." Observers also coded the object of the child's attention (environment, other child, group of children, or adult) and the duration of the child's activities. (P. 24)

In addition, all children were administered two standard tests of intellectual development: Caldwell's Preschool Inventory (PSI) and a modified version of the Peabody Picture Vocabulary Test (PPVT). The following summary of major findings to date is based on a preliminary report (1978) supplemented by personal communications from the study directors regarding results of more recent analyses.

With respect to the stated objective of assessing the developmental impact of caretaker-child ratio and group size, the results were qualified by the age of the child. In a special substudy of center care arrangements for infants under three, the investigators found that the caretaker-child ratio was more important than group size in affecting the behavior of caretakers and also of children, although observations of the latter were more limited. The more infants there were per staff member, the less time the caretaker spent in teaching—either formal or informal—and the more she engaged in management and control behavior or simply observation. Infants in a low adult-child ratio situation were more likely to exhibit distress reactions or to be apathetic and passive. Increased group size had similar effects but of much smaller magnitude (Connell, personal communication).

In group care programs enrolling children from three to five, the more critical factor was group size. In their preliminary report,

Travers and Ruopp summarize their findings as follows: "At this point, it is clear that groups of 15 or fewer children, with correspondingly small numbers of caregivers, are associated with higher frequencies of desirable child and caregiver behavior and higher gains on the PSI and PPVT than groups of 25 or more children" (p. 35).

Although caretaker-child ratio did show some significant relationships to outcome measures, most of these effects became negligible after group size was introduced as a control variable (J. Travers, personal communication). Holding the child-staff ratio constant, however, did not eliminate the effect of group size: "For example, groups of 12–14 children with two caregivers had, on the average, better outcomes than groups of 24–28 children with four caregivers. These results make it clear that, staff/child ratio, cannot by itself be the principal mechanism for guaranteeing benefits to children, although it may be an important indicator of staff burden ... The effects of staff/child ratio were minor when compared with those of group size" (p. 36).

Even when the effects of staff-child ratio were tested experimentally, the results were marginal at best. In a separate phase of the study using eight centers in the Atlanta public schools, children were assigned randomly to classes with high (1:5.5) and low |1:7.8) ratios. A reliable treatment effect was found in gain scores on one test of intellectual performance (PSI) but not on the other (PPVT), and even the significant association obtained was much weaker than that between group size and PSI score in the main study.[2]

What is it about group size that makes the difference? More broadly, what is "the principal mechanism for guaranteeing benefits to children"? Our first evidence is in the form of observational data from the main Abt study, documenting differences in the behavior of three- to five-year-olds and their caretakers as a function of size of group. The investigators describe the patterns as follows:

Caregiver Behavior: Lead teachers in smaller groups engaged in more social interaction with children (questioning, responding, instructing, praising and comforting) than did teachers in larger groups. In contrast, teachers in larger groups spent more time observing children and interacting with other adults than did teachers in smaller groups . . .

Child Behavior: Children in smaller groups showed higher frequencies of such behaviors as considering/contemplating, contributing ideas, giving opinions, persisting at tasks and cooperating than did children in large groups. In general, smaller groups were characterized by high levels of interest and participation on the part of children. In large groups, children showed higher frequencies of wandering, non-involvement, apathy and withdrawal. (Pp. 36–37)

It will be observed that the behaviors engaged in by the teachers in smaller groups were of the kind that stimulated, sustained, and encouraged task-oriented and cooperative activities on the part of the children. The children also differed in their performance on tests of intellectual performance, with those cared for in smaller groups showing significantly higher gains during the school year.

One additional analysis conducted by the investigators focuses even more sharply on the crucial structural feature giving rise to the observed patterns. Rather than using group size as defined by the number of children and adults present in the setting, the researchers employed an index based on the number of persons actually interacting with each other. When this functional measure of group size was used as an independent variable, the correlations with outcome measures significantly increased (J. Travers, personal communication).

These findings accord nicely with our hypotheses regarding the importance to the child's development of involvement with an adult in progressively more complex patterns of reciprocal molar activities (hypotheses 1 through 7). What the results of the Abt study show is that, in day care centers for children between three and five years of age, such task-oriented activities, as distinguished from management and control activities, are more likely to occur as the group size becomes smaller.

If task-oriented interaction is the key, why is it better predicted by group size among older preschoolers than by the caretaker-child ratio for infants under three? The available data permit no definitive answers to this question but do suggest some plausible explanations. It is important first to recognize that caretaker-child ratios for infant day care are substantially higher and group sizes considerably smaller than those for older preschoolers. In their special substudy of fifty-four day care centers for children under three, Travers and Ruopp report that observed staff-child ratios were higher than state-required minimums, averaging 1 to 3.8 for babies under eighteen months compared with a modal requirement of 1 to 5.3. The corresponding ratios for toddlers (eighteen to thirty months) showed a similar pattern but were somewhat lower (1 to 6.1 versus 1 to 7.8). Conversely, groups were smaller for infants under eighteen months of age than for toddlers, averaging 6.9 versus 10.9 children.

Under these circumstances it seems likely that, in centers for infants under three, the addition of another staff member would do more to increase opportunities for reciprocal caregiver-child activity than a reduction in group size. Conversely, in centers for three- to five-year-olds, with typical group sizes of ten to twenty-five or more and only three or four caregivers at most, the occurrence and longevity of joint activity dyads are more likely to be a function of the number of children present than of the adult-child ratio. Unfortunately, I know of no research data bearing directly on this issue.

Along with the purely numerical factors there are substantive considerations pertaining to these variables which are relevant to the child's developmental status. The caretaker-child ratio takes into account the presence of adults, whereas the variable of group size does not. We have already reviewed research evidence (in chapters 4 and 7) documenting the importance of a one-to-one relationship between infant and adult in maintaining the young child's emotional security and enabling him to explore and learn from the immediate environment. It is consistent with this line of evidence that a low caretaker-child ratio should be associated with greater distress and apathy among infants under three years of age. During this early period, age-mates, as compared with adults, play a relatively minor role in the child's development, and it is only afterwards that the peer group becomes a potent force in the lives of young children (Hartup, 1970). Hence the number of age-mates present in the setting is not likely to have much significance for the infant before the age of three. After that time, however, not only do peers exercise an increasingly powerful influence, but the child's dependence on a one-to-one relationship with an adult markedly diminishes, and he becomes able to function effectively and to learn in somewhat larger groups (provided they are not so large as to reduce below a critical level the occurrence of developmentally effective interactions between a child and an adult). The shifting pattern of influence reported by the Abt investigators is in agreement with these developmental facts.

What is more, the observed pattern is consistent with a more specific generalization drawn from our review of research on the effects of group care: the exposure of the child to group experience with peers tends, at least in contemporary American society, to undermine the socialization efforts of adults and to invite the emergence of egocentrism, aggression, and antisocial behavior. Once children are beyond the age of three, it is reasonable to expect that the larger the peer group, the weaker will be the influence of the supervising adult. As the child approaches school age, group size can act as a catalyst in shifting the balance of power from adults to peers with corresponding impairment of developmental progress.

If this analysis is correct, we are presented with a paradox in the face of the New York day care study findings (Golden et al., 1978). It will be recalled that, on the basis of observations conducted in both types of settings, family day care programs were found to be "superior to the Group Day Care programs in the caregiver-child ratio, in the amount of social interaction and individual attention children receive from caregivers, and in the degree of positive social-emotional stimulation provided by caregivers to children during the noon meal" (p. 148). While no data are reported on group size, there can be little doubt that the average number of children in a family day care home was smaller than in a center.

On the basis of findings from the national day care study, one might therefore expect family day care to produce more developmental gains than center care. Yet the New York City study found the very opposite. By three years of age, the children enrolled in group care obtained significantly higher scores on the Stanford Binet than their matched controls enrolled in family day care (IQs of 99 versus 92). Moreover, whereas the children in center care had maintained the same level of intellectual performance between eighteen and thirty-six months of age, their counterparts in family day care had shown a significant decline from ninety-eight to ninety-two.

How is the paradox to be resolved? We begin by recalling that, in the national day care study, the changes associated with reduction in group size involved not only the amount of staff-child interaction but its content as well. As group size decreased, caretakers engaged in more "questioning, responding, instructing, praising and comforting" (Travers and Ruopp, p. 36). The only reliable differences in caregiver behavior by setting found in the New York City study were in the frequency of adult-child interaction without any regard to content and in the amount of interaction and positive socioemotional stimulation directed to children during the noon meal (Golden et al., p. 144). Both of these differences favored the family day care group. But the sole substantive measure of caregiver activity, "cognitive language stimulation," did not show a significant effect by setting; the direction of the obtained difference is not given. Taking into account the nature of the caregiver variables on which centers were surpassed by family day care settings, there is no longer any necessary contradiction between the results of the New York City and the national studies. It appears that the critical variable is the content of the interactions rather than the amount. But what of the higher caretaker-child ratio found in the family day care homes? Does not that run counter to the trend reported by Travers and Ruopp of more positive outcomes in centers with higher staff-child ratios, especially in centers serving infants under three? The question receives an affirmative answer only if one assumes that this ratio has the same effect regardless of the content of adult-child interactions.

An alternative hypothesis focuses attention on yet another element of the microsystem: how the staff members in each setting view their role, that is, to what extent the caregiver in family day care is perceived by himself and others not only as caring for the child and playing with him but also as engaging in formal and informal teaching. We have already seen powerful evidence that, when such differential perceptions exist, they are likely to be implemented in actual behavior. This phenomenon is what may have occurred in the New York City and national day care studies and, if so, could explain their paradoxical findings. By contrast with family day care homes, centers which in their physical and social characteristics begin to approach preschools and schools, tend to evoke more teacherlike behavior, both formal and informal. In the smaller group settings, where the caregiver is more likely to become involved in face-to-face interaction with the child, he is therefore more apt to engage in "questioning, responding, and instructing" along with the "praising and comforting." And when he does so, the children respond with "higher frequencies of such behaviors as considering, contemplating, contributing ideas, giving opinions, persisting at tasks and cooperating" (Travers and Ruopp, p. 36).

Lest we jump to any premature conclusons about the superiority of the day care center to the family day care home as a context for development, we need to remind ourselves of the inferential nature of the argument. The contrasting pattern was found in a comparison not of day care centers and day care homes but of large and small groups in center settings. Regrettably, the specific types of task-oriented child activities observed in the national day care study were not examined in the New York City project. The closest approximation was an "Index of Child's Cognitive Language Behavior." Family day care children obtained significantly higher scores on this measure at one year of age than their counterparts in center care but did not differ in subsequent assessments at ages one and a half and two.

Moreover, as previously noted, it is difficult to assess the developmental significance of such differences, or lack thereof, when children are compared in quite different situations. In view of this fact, it is unfortunate that neither the New York City nor the national day care studies included observations in the children's own homes.

Apart from the absence of directly comparable data, there are even more compelling grounds for caution in inferring developmental superiority for group care as a function of more educationally oriented activity on the part of center caregivers. We need only recall that adults are not the only influential figures in the center setting; age-mates are present as well and in much greater number. We have already documented the tendency of peers to undermine the socialization efforts of supervising adults and to invite the emergence of egocentrism, aggression, and antisocial behavior. In the New York City study, the only variable on which children in family day care consistently surpassed their center counterparts was interaction and social competence with adults.

More definitive data are needed to specifiy the differential impact of adults and peers in family day care and center settings. Such data must include information not merely about the frequency but also about the content of adult-child and peer-child activities in the day care settings, and similarly substantive outcome measures need to be obtained from both groups of children observed with their families at home and, later on, in school. Only by employing such transcontextual, mesosystem designs and assessing the emerging patterns of molar activity, can the particular ecological properties of day care environments that affect the course of the child's development be identified.

There is no investigation of preschools that analyzes the impact of particular program components on the observed behavior of both teachers and children. The only systematic studies of program variations within a single research design compare total projects employing different curricula, using test scores as outcome measures (Di-Lorenzo, 1969; Karnes, 1969; Soar, 1972). The general finding is that the more structured, cognitively oriented programs produce larger gains that are somewhat more enduring. There are some indications, however, that highly structured curricula may have some less commendable side effects outside the sphere of academic achievement. Bissell (1971), in an analysis of results from a national research program evaluating different approaches in the Head Start program, found that children enrolled in more structured programs were more likely to give passive responses on the Hertzig-Birch (Hertzig et al., 1968) measures of coping style. According to Bissell, the results suggest that the children have learned what a question is and what an appropriate answer is. Such an orientation may be far more adaptive to the kinds of tasks required of the child in the primary grades than to the expectations of intellectual initiative in defining and solving problems encountered in the upper grades.

In the same vein, analysis of data from Follow-Through classrooms (Stanford Research Institute, 1971a, 1971b) indicated that changes in attitude toward school and learning were more likely to occur in the "Discovery" approaches than in the "Structured Academic" curriula, although it was children enrolled in the latter programs who made particularly large gains. Moreover, in the "Discovery" groups, there was a strong association between positive shifts in attitudes toward school and gains in achievement. No such relation obtained in the "Structured Academic" approaches. The Soars have demonstrated that greater amounts of academic growth over the summer were associated with an unstructured individual teaching style during the preceding school year rather than with a structured, direct style (Soar, 1966; Soar and Soar, 1969).

Such findings undescore the importance of using more differentiated, ecologically oriented measures of both program characteristics and outcomes to arrive at an understanding of the origins or the developmental effects of preschool and school experience.

A line of evidence emerging from my (1974d) comparison of preschool projects employing different intervention strategies calls attention to possible limitations of even enriched preschool environments as a context for human development. The programs were selected to reflect the principal approaches currently in use and were of four major types: *group programs* conducted in preschool settings—these were all well funded and university based with trained staff and high teacher-child ratios; *home-based parent-child intervention* implemented by a trained home visitor who demonstrated and encouraged developmentally stimulating activities to be engaged in jointly by mother and child; *home-based tutoring* carried out by a home visitor with the child but not involving the parent; and *preschool-home combination,* in which parent-child intervention in the home was also provided to children simultaneously enrolled in preschool.

Children in all four types of programs showed substantial gains in intelligence test scores, but when intervention was conducted only in the home and involved mother and child jointly, these gains persisted longer after the program was discontinued. In addition, the involvement of the mother resulted in diffusion effects to younger siblings. The mothers themselves were also positively affected. They began to show more self-confidence and to exhibit successful initiative and accomplishment in educational, occupational, and community activities.

Two elements were identified as critical to the success of so-called parent-child intervention. The first was "the involvement of parent and child in verbal interactions around a cognitively challenging task" (p. 54). The second was insuring and reinforcing the parent's status as the key person in the child's life.

The importance of each of these factors is illustrated by the outcomes of programs in which one or the other condition was not met. Thus in Schaefer's home-based tutoring project (Schaefer, 1968, 1970; Schaefer and Aaronson, 1972), where the home visitor worked with the young child but not with the mother, the experimental effects began to disappear even while the program was still in operation. The need to establish and maintain the parent's status as the central figure in the child's life is reflected in the outcome of an experimental program conducted by Karnes and her colleagues (1969). Encouraged by the results of the mother-intervention program, the researchers thought of achieving a still better outcome by combining it with a preschool experience for the children themselves. To maximize the joint effect, the intervention workers "made a major effort to coordinate the teaching efforts at home with those at school" (p. 205).

IQ gains achieved over a two-year period were compared with those obtained in other, similarly selected preschool classes enrolling pupils whose mothers did not participate in a special program. Given the success achieved previously with children of the same age by a program involving mother-intervention only, the results of the combined strategy came as a disappointing surprise. The fourteen-point again in IQ made by the control group of children was actually larger than the twelve-point rise achieved by the experimental group, although the difference was nonsignificant. The preschool-only group did score reliably higher in tests of language development.

Why did the mother-intervention program fail to make any added contribution? In the judgment of the investigators, the explanation lies in certain changes that occurred in the home-based program as the result of combining it with the preschool.

These changes, which seemed relatively minor at the time, coupled with the child's preschool attendance may have significantly altered the mother's perception of her role in this program. In the [previous] study, the mother was aware that she was the only active agent for change in her child, and as she became convinced of the merit of the program, she increasingly felt this responsibility. The fact that project staff placed a similar value on her role was demonstrated to the mother by the weekly checklist and the biweekly home visits to evaluate her work. In the [present] study, mothers appreciated the value of the activities for their children but may have overemphasized the role of the preschool in achieving the goals of the program. Teachers, through their actions rather than direct statement, may have unwittingly reinforced this devaluation of mother-child interaction by making the purpose of home visits the delivery of materials to absentee mothers. The emphasis of home visits had changed from concern over mother-child interaction to concern over the presence of materials, and it was not unreasonable for some mothers to feel that the materials themselves were the essential ingredient in effecting change. Through the weekly checklist the *mother* had reported what she taught *at home,* but during the three visits made in conjunction with the operation of the preschool, the *teacher* reported on the progress of the child *at school.*

Mothers in the [previous] study saw the major intent of the program to be the benefits which fell to their children. In the [present] study, since the children already received the benefits of a preschool experience, the mothers tended to use the mother-involvement program to meet personal needs. Instead of a mother's program *for children,* the program may have been seen as a mother's program *for mothers.* Evaluations of the . . . program, both verbal and written from teachers and mothers, support this view. Mothers frequently commented on their enjoyment of the social aspects of the program and on the genuine pleasure they experienced in making educational materials for their children, but a disturbing number of mothers also indicated at the end of the year that the primary use of these materials at home was by the child alone or under the direction of older siblings. Apparently mothers felt that they had fulfilled their responsibility to the program when ·they sent their children to school, attended a weekly meeting, and made educational materials, and, indeed, this level of involvement represented a major commitment. To some extent, mothers may have substituted these experiences for direct mother-child interaction, a consequence counter to the intent of the study, and that substitution may have been detrimental to the development of verbal expressive abilities. The solitary involvement of a child with the materials or their use with a sibling not trained to encourage verbal responses is consistent with such a performance. (Pp. 211–212)

These findings regarding the effects on parent-child intervention at home conducted either separately or jointly with a center-based preschool program lead to both familiar and novel conclusions. They provide corroboration from yet another source for our basic hypotheses regarding the conditions most conducive to psychological growth. The results underscore the importance of the child's involvement in progressively more complex, joint, reciprocal activities in the context of a primary dyad (hypotheses 1 through 7), they testify to the power of third parties in enhancing or impairing the capacity of a primary dyad to function effectively as a developmental context (hypothesis 8), and they illustrate the significance of the role status and power accorded to the caregiver (hypotheses 9 through 11) as a function of "the existence of other roles in the setting that invite or inhibit behavior associated with the given" (hypothesis 12).

The findings of research on parent-child intervention also introduce a new element of both theoretical and practical significance. Thus far, the third parties and roles that have been the principal focus of our attention have all come from the same setting (spouse, sibling, wardmate, and so on). In parent-child intervention, however, the third party comes from the outside as a representative of another and different setting, in this instance a university-based or agency-sponsored program. This relation signals the involvement of the mesosystem, which concerns the interconnections between settings. Yet the same principles seem to apply as for the microsystem.

The studies we have examined reveal that group settings for the care and education of young children differ from homes primarily in the nature of the molar activities in which adults and children engage and in the extent and character of relations that develop between children and adults.

HYPOTHESIS 20
The immediate and long-range effects of exposure to group settings in early childhood will be reflected not primarily in scores on intelligence, achievement tests, or interaction processes but in the nature and variety of the molar activities engaged in by the child and in the changed character of his behavior and relations toward adults and peers.

The research also shows that group settings for young children have the capacity to enhance the development of intellectual and educational competence during the preschool years and after the child has entered school. The results of these investigations further indicate that the power of preschool environments to produce these immediate and longer-range effects are primarily a function of their distinctive ecological characteristics as set forth in the preceding hypothesis.

HYPOTHESIS 21

The capacity of group settings for young children to enhance development of intellectual and educational competence depends on the extent to which caregivers and preschool personnel, in their interactions with children, engage in behaviors that stimulate, sustain, and encourage task-oriented activities on the part of the child. Examples of such adult behaviors include questioning, instructing, responding, praising, and comforting. The more often adults exhibit behaviors of this kind, the more the children become capable of task-oriented and cooperative activities (such as persisting in tasks, thinking, contributing ideas, giving opinions, and working together).

HYPOTHESIS 22

The ability of caregivers or preschool teachers to engage in activities that facilitate the children's development is a function of setting properties that vary with the age of the child. In settings for infants under three years, where group sizes are relatively small, adult-child ratio becomes a critical factor in influencing the ability of caretakers to engage in the kind of reciprocal, one-to-one interaction that appears to be most effective in meeting the needs and facilitating the development of the very young child. In settings for children between three and five, where the number of children under care is large, the size of the class—more specifically of the functional group—becomes a major determiner of both caretaker and child activities. Larger groups not only reduce the frequency of developmentally effective activity on the part of the adults but also increase the possibility of children's remaining uninvolved or becoming disengaged, or caught up in tangential or counterproductive diversions with their age-mates.

As the last part of the hypothesis implies, along with their capacity to sustain and enchance task-oriented activity and intellectual competence, group settings for young children can have effects that are regressive from the perspective of goals for socialization prevailing in the society at large.

HYPOTHESIS 23
Children who from an early age are cared for in group settings
for most of the day are more likely to engage in egocentric,
aggressive, and antisocial behavior both during the preschool
years and through later childhood into adolescence. The observed
effect is particularly marked for boys. It is mediated through the
children's peer group and is most likely to occur in societies that
encourage the expression of individualism, aggression, and independence in children's groups, especially by boys.

To conclude that the hypotheses we have been able to derive
from existing research constitute. all the basic ecological processes
operating in day care and preschool settings is seriously to underestimate the power of these environments to influence psychological
growth. If the ecological principles emerging from our theory are
valid, then day care and preschool experiences can have much more
impact than is indicated by the results of the kinds of investigations that have been conducted thus far. The underestimation of
day care and preschool effects derives from the limitations of the
conventional research model employed in almost all the studies we
have examined. These limitations appear in four domains.

1. In previous studies, little systematic attention has been accorded to an examination of the variety and complexity of molar
activities as manifestations of both development in the person and
the developmental potential of the setting in which the person is
found. The issue may be stated in the form of a hypothesis susceptible to empirical test.

HYPOTHESIS 24
The variety and complexity of the molar activities available to and
engaged in by the child in a day care or preschool setting affects
her development as manifested by the variety and complexity
of the molar activities exhibited by the child in other settings,
such as the home and, subsequently, the school.

This hypothesis would be most efficiently investigated in a before-
and-after design focusing on the ecological transition of the child
from home to a preschool setting or from a preschool setting to
school. In both instances, observations would be focused on changes
in molar activities in the home following the child's entry into the
external group setting. In this second case, observations would also
be conducted in the preschool center and the school classroom and
involve a comparison of two groups, one with and one without prior

experience in a preschool setting. The prediction from the hypothesis is that, by contrast with the results likely to be obtained using traditional psychometric measures or observations of interaction processes, the analysis of the *content* of *molar* activities would reveal substantial differences reflecting the developmental impact of experience in one setting on behavior and subsequent development in another.

2. In prior investigations, little systematic attention has been given to the nature and complexity of interpersonal structures either as manifestations of the development of the person engaging in these subsystems or as indicators of the development potential of the setting in which they occur. The nature of the interpersonal structure is defined by the pattern of reciprocity, balance of power, and affective relation exhibited in the constituent dyads; the degree of complexity is reflected in the magnitude of the $N + 1$ system involved (dyad, triad, and so on). Again, the issue can be formulated as a testable hypothesis.

HYPOTHESIS 25
The nature and complexity of the interpersonal structures available to and engaged in by the child in a day care or preschool setting affects her development as manifested by the nature and complexity of the interpersonal structures initiated or entered into by the child in other settings, such as the home and, subsequently, the school.

The design most appropriate for the investigation of this hypothesis is the same as that proposed for the preceding one, but now with a primary focus on interpersonal structures. As before, the prediction is that the child's experience in a day care or preschool environment will have substantial impact on the kinds of structures in which she participates in other settings.

3. Since previous studies have accorded little systematic attention to either molar activities or interpersonal structures, they have not been concerned with the environmental conditions that facilitate or impair the occurrence of these phenomena. The next hypothesis specifies circumstances within the setting itself that are relevant in this regard.

HYPOTHESIS 26
The developmental potential of a day care or preschool setting depends on the extent to which supervising adults create and

maintain opportunities for the involvement of children in a variety of progressively more complex molar activities and interpersonal structures that are commensurate with the child's evolving capacities and allow her sufficient balance of power to introduce innovations of her own.

4. Prior studies of development in day care and preschool environments have concentrated almost exclusively on events within the setting rather than on the interconnections between that setting and others in which the child spends her time. In the next chapter, I argue that the capacity of any setting—like the day care center, the preschool center, or for that matter the home—to generate and sustain ongoing molar activities and stable interpersonal structures depends on the relationships between that setting and others.

Finally, it is important to note that the hypotheses developed in this chapter are not limited in their applicability to day care and preschool settings but extend to classrooms, playgrounds, camps, and other environments in which children live and grow.

PART FOUR

Beyond
the Microsystem

9.

The Mesosystem
and Human Development

In analyzing the forces that affect processes of socialization and development at the level of the mesosystem, we shall find ourselves using most of the same concepts employed to delineate the structure and operation of microsystems. Thus the basic building blocks will be the familiar elements of the setting: molar activities, roles, and interpersonal structures in the form of dyads and $N+2$ systems varying in the degree of reciprocity, balance of power, and affective relations. What is more, many of the hypotheses derived will be analogous to prototypes previously formulated for the microsystem. The difference lies in the nature of the interconnections involved. At the microsystem level, the dyads and $N+2$ systems, the role transactions, and the molar activities all occur within one setting, whereas in the mesosystem these processes take place across setting boundaries. As a result of this isomorphism, it is possible to formulate most of our hypotheses in advance and then examine relevant research evidence.

I have defined the mesosystem as a set of interrelations between two or more settings in which the developing person becomes an active participant. What kinds of interconnections are possible, for example, between home and school? I propose four general types.

1. *Multisetting participation.* This is the most basic form of interconnection between two settings, since at least one manifestation of it is required for a mesosystem. It occurs when the same person engages in activities in more than one setting, for example, when a child spends time both at home and at the day care center. Since such participation necessarily occurs sequentially, multisetting participation can also be defined as the existence of a direct or *first-order* social network across settings in which the developing person is a participant. The existence of such a network, and therefore of

a mesosystem, is established at the point when the developing person first enters a new setting. When this occurs, we also have an instance of what I have called an *ecological transition,* in this instance a transition from one setting to another.

When the developing person participates in more than one setting of a mesosystem, she is referred to as a *primary link,* as when Mary enters school. Other persons who participate in the same two settings are referred to as *supplementary links*; for instance, Mary's mother attends a PTA meeting, her teacher pays a visit to the home, or Mary brings home a classmate to play. As these examples indicate, direct links can operate in the direction of either setting.

A dyad in either setting that involves a linking person as a member is referred to as a *linking dyad.*

2. *Indirect linkage.* When the same person does not actively participate in both settings, a connection between the two may still be established through a third party who serves as an *intermediate link* between persons in the two settings. In this case, participants in the two settings are no longer meeting face-to-face so that we speak of them as members of a *second-order network* between settings. Such second-order connections can also be more remote, involving two or more intermediate links in the network chain.

3. *Intersetting communications.* These are messages transmitted from one setting to the other with the express intent of providing specific information to persons in the other setting. The communication can occur in a variety of ways: directly through face-to-face interaction, telephone conversations, correspondence and other written messages, notices or announcements, or indirectly via chains in the social network. The communication may be one-sided or may occur in both directions.

4. *Intersetting knowledge* refers to information or experience that exists in one setting about the other. Such knowledge may be obtained through intersetting communication or from sources external to the particular settings involved, for example, from library books.

The most critical direct link between two settings is the one that establishes the existence of a mesosystem in the first instance—the *setting transition* that occurs when the person enters a new environment. If the child goes to school on the first day unaccompanied, and no one else from his home enters the school setting, there exists only a single direct link between the two microsystems. Under these circumstances, the transition and the resulting link that is established are referred to as *solitary*. Should the child be accompanied by his mother or an older brother who enters the school with him and introduces him to the teacher or to the other children, the transition and the resultant link are described as *dual*. Of course the mother may not come to the school until a later point, or the teacher may visit the home, in which case the connection becomes dual at that time. A mesosystem in which there is more than one person who is active in both settings is referred to as *multiply linked*. A mesosystem in which the only links, apart from the original link involving the person, are indirect or in which there are no additional links whatsoever is described as *weakly linked*.

I make these distinctions not merely because they are logically possible but because I believe them to be of significance for the way in which the developing person is able to function in new settings. A dual transition permits the formation of a three-person system immediately upon entry into the new setting, with all its potential for second-order effects; the third party can serve as a source of security, provide a model of social interaction, reinforce the developing person's initiative, and so on. The extent of this catalytic power of the intermediary depends on his relation with the developing person as well as on the nature of the dyads established in the new setting, that is, whether they are only observational (the mother acts purely as a visitor), involve joint activity (the mother converses with the teacher), or develop into a primary dyad (the mother and teacher become good friends).

These considerations are made explicit in two sets of hypotheses. The first set focuses on the experience of the developing person in the mesosystem; these hypotheses deal with the structure of primary links and their developmental consequences. The second series is concerned with analogous considerations pertaining to supplementary links. We begin with hypotheses that specify optimal conditions for the establishment and maintenance of the primary link.

HYPOTHESIS 27
The developmental potential of a setting in a mesosystem is enhanced if the person's initial transition into that setting is not made alone, that is, if he enters the new setting in the company of one or more persons with whom he has participated in prior settings (for example, the mother accompanies the child to school).

HYPOTHESIS 28
The developmental potential of settings in a mesosystem is enhanced if the role demands in the different settings are compatible and if the roles, activities, and dyads in which the developing person engages encourage the development of mutual trust, a positive orientation, goal consensus between settings, and an evolving balance of power in favor of the developing person.

To consider a negative example: as indicated by the results of a pilot study conducted by me and my colleagues (Avgar, Bronfenbrenner, and Henderson, 1977; Cochran and Bronfenbrenner, 1978), mothers from two-parent families who hold part time jobs find themselves in a difficult role conflict; the husbands continue to act as if their wives were still functioning as full time mothers, while employers often treat them as if they were full time employees. The mothers experience the resulting frustration as impairing their effectiveness as parents, their performance on the job, and their development as human beings.

Thus participation in more than one setting has developmental consequences. From infancy onward, the number of settings in which the growing person becomes active gradually increases. This evolving participation in multiple settings is not only a result of development—under certain conditions, it is also a cause. This thought is developed in a series of hypotheses.

HYPOTHESIS 29
Development is enhanced as a direct function of the number of structurally different settings in which the developing person participates in a variety of joint activities and primary dyads with others, particularly when these others are more mature or experienced.

Based on this hypothesis one could make the following prediction: holding age and socioeconomic factors constant, a young person entering college who has been closely associated with adults outside the family, has lived away from home, and held a number of jobs will be able to profit more from a college education than one whose experience has been more limited.

The hypothesis is based on the assumption that involvement in joint activity in a range of settings requires the developing person to adapt to a variety of people, tasks, and situations, thus increasing the scope and flexibility of his cognitive competence and social skills. Moreover, as indicated earlier, joint activities tend to develop a motivational momentum of their own that persists when the participants are no longer together. When such activities occur in a variety of settings, this motivational momentum tends to generalize across situations. These effects are further enhanced if the participants are emotionally significant in each other's lives, that is, if they are members of primary dyads. The hypothesis has a corollary at the sociological level.

HYPOTHESIS 30
The positive developmental effects of participation in multiple settings are enhanced when the settings occur in cultural or subcultural contexts that are different from each other, in terms of ethnicity, social class, religion, age group, or other background factors.

Underlying this hypothesis is the assumption that differences in activities, roles, and relations are maximized when settings occur in culturally diverse environments.

A critical case for the two foregoing hypotheses would be represented by a person who had grown up in two cultures, had participated actively and widely in each society, and had developed close friendships with people in both. If the two hypotheses are valid, such a person, when compared with someone of the same age and status who had grown up in only one country and subculture, should exhibit higher levels of cognitive function and social skill and be able to profit more from experience in an educational setting. I know of no research on this phenomenon, but it is certainly susceptible to empirical investigation by, for example, comparing the development of children with and without extensive experience of other cultures or ethnic groups, holding other aspects of family background constant. The hypotheses could also be tested by, for instance, assigning youngsters to work projects involving participation in subcultures within the community.

This line of reasoning is applicable not only at the level of the individual but also at that of the dyad. Just as it is possible for a person to engage in activity in more than one setting, so can the dyad do this. Such a migrating two-person system is referred to as a *transcontextual dyad*. From an ecological perspective, there is reason to expect this type of structure to have special significance for development. It is probably even more conducive to the formation of primary dyads than a joint activity limited to a single setting. But more important, I suggest that the occurrence of transcontextual dyads in the life of the person may operate to enhance the person's capacity and motivation to learn. This possibility is based on the assumption that when a variety of joint activities are carried out in a range of situations but in the context of an enduring interpersonal relationship, the latter both encourages the development of higher levels of skill and tends to generate especially strong and persistent levels of motivation. This thinking leads to the following three hypotheses.

HYPOTHESIS 31
The capacity of the person to profit from a developmental experience will vary directly as a function of the number of transcontextual dyads, across a variety of settings, in which she has participated prior to that experience.

HYPOTHESIS 32
Children from cultural backgrounds that encourage the formation and maintenance of transcontextual dyads are more likely to profit from new developmental experiences.

HYPOTHESIS 33
Development is enhanced by providing experiences that allow for the formation and maintenance of transcontextual dyads across a variety of settings.

Several hypotheses pertain to the optimal structure of additional links between settings beyond the primary connection established by the developing person. The first one is merely an extension of an earlier hypothesis (28) now expanded to encompass any additional persons who participate in the different settings under consideration.

HYPOTHESIS 34
The developmental potential of settings in a mesosystem is enhanced if the roles, activities, and dyads in which the linking person engages in the two settings encourage the growth of mutual trust, positive orientation, goal consensus between settings and an evolving balance of power responsive to action in behalf of the developing person. A supplementary link that meets these conditions is referred to as a *supportive link*.

An example in which the conditions stipulated in this hypothesis were violated is found in the previously cited (chapter 8) account by Karnes of the unforeseen effects of combining home visits with a preschool program. The change in the staff's treatment of mothers as a result of the new arrangement decreased the mother's sense of her own importance and efficacy and her active involvement as as key figure in her child's development.

HYPOTHESIS 35
The developmental potential of a setting is increased as a function of the number of supportive links existing between that setting and other settings (such as home and family). Thus the least favorable condition for development is one in which supplementary links are either nonsupportive or completely absent—when the mesosystem is weakly linked.

HYPOTHESIS 36
The developmental potential of a setting is enhanced when the supportive links consist of others with whom the developing person has developed a primary dyad (the child's father visits the day care center) and who engage in joint activity and primary dyads with members of the new setting (the child's mother and teacher are bridge partners).

The examples in hypothesis 36 assume that parents behave, as indeed they usually do, in a manner consistent with the requirements for a supportive link as stipulated in hypothesis 34.

Our next hypothesis in effect sets a boundary condition to the relationships posited in the three preceding ones.

HYPOTHESIS 37
The relationships posited in hypotheses 34 through 36 vary inversely with the developing person's prior experience and sense of competence in the settings involved. Thus the positive impact of linkage would be maximal for young children, minorities (especially in a majority milieu), the sick, the aged, and so on. Conversely, as experience and self-confidence increase, the postulated relationships would decrease in magnitude to a point at which they reverse direction, such that for a maturing person who is at home in her own culture, development may be further enhanced by entry into new settings that have no prior links with the setting of origin or in which the balance of power is weighted against the developing person and those operating in her behalf.

In other words, the hypothesized relationships are curvilinear with a turning point that depends on the person's stage of development and social status in the society. For a young teen-ager leaving home for the first time—or a minority member visiting city hall— going with a friend or knowing someone in the new location can make a difference. For a successful college graduate, looking for a job in a new environment might be more conducive to development than staying at home to work in the family business.

Second-order social networks involving intermediate links can perform at least three important functions. They provide an indirect channel for desired communication in situations where no direct link is available. (For example, a working mother who cannot attend parents' meetings at the day care center can find out what happened from a friend). Second-order networks can also be used for identifying human or material resources from one setting needed for use in the other. (For instance, a parent turns to friends for help in finding a job.) Perhaps the most important mesosystem function of social networks is unintended: they serve as channels for transmitting information or attitudes about one setting to the other. (From third parties, parents can be told a different story about what happened at school from the version brought home by the child, or the teacher can learn "via the grapevine" that parents are prejudiced against her because of her ethnic or religious background).

Our hypothesis specifying the structure of indirect links most conducive to development follows a familiar pattern, one that defines a supportive function for interconnections between settings.

HYPOTHESIS 38
The developmental potential of a mesosystem is enhanced to the extent that there exist indirect linkages between settings that encourage the growth of mutual trust, positive orientation, goal consensus, and a balance of power responsive to action in behalf of the developing person.

We have already noted that intended communication between settings can take a variety of forms and can vary in the direction of flow. These are parameters that have been extensively investigated in communication research. I have drawn on this literature to derive three general hypotheses addressing the influence of communication between settings on their potential as contexts for development.

HYPOTHESIS 39
The developmental potential of participation in multiple settings will vary directly with the ease and extent of twoway communica-

tion between those settings. Of key importance in this regard is the inclusion of the family in the communications network (for example, the child's development in both family and school is facilitated by the existence of open channels of communication in both directions).

HYPOTHESIS 40
The developmental potential of settings is enhanced to the extent that the mode of communication between them is personal (thus in descending order: face-to-face, personal letter or note, phone, business letter, announcement).

Information available in one setting about another can come from a variety of sources. Besides direct oral and written communication between settings these can include traditional knowledge handed from one generation to the next, one's own childhood experience, books, television, and so on (Lüscher and Fisch, 1977). Especially important are discussions that take place in one setting about the other. For example, parents of a young child can describe to him what school will be like, or the school can offer courses in family life. Thus intersetting knowledge also takes a variety of forms. In addition to oral or written information, advice, and opinion, it may involve objects from or representing the other setting (as when a child takes a favorite toy to show at school or a school banner hangs in the child's bedroom) as well as experiences, both imaginary (such as role playing) and real (such as introductory visits).

As these examples indicate, intersetting knowledge can serve two somewhat different functions identified in the following two hypotheses.

HYPOTHESIS 41
Development is enhanced to the extent that, prior to each entry into a new setting (for instance, enrolling in day care or school, being promoted, going to camp, taking a job, moving, or retiring), the person and members of both settings involved are provided with information, advice, and experience relevant to the impending transition.

HYPOTHESIS 42
Upon entering a new setting, the person's development is enhanced to the extent that valid information, advice, and experience relevant to one setting are made available, on a continuing basis, to the other.

It would be interesting to examine how the hypotheses dealing with intersetting connections apply to a specific situation: the developmental impact on the child of relations between the home and group settings such as day care, preschool, and school. If our hypotheses are valid, we would expect research on this question to reveal more advanced development for children growing up in environments characterized by certain types of interconnections between the home and, for example, the school. These interconnections would be characterized by more frequent interaction between parents and school personnel, a greater number of persons known in common by members of the two settings, and more frequent communications between home and school, more information in each setting about the other, but always with the proviso that such interconnections not undermine the motivation and capacity of those persons who deal directly with the child to act in his behalf. This qualification gives negative weight to actions by school personnel that degrade parents or to parental demands that undermine the professional morale or effectiveness of the teacher. Analogous considerations would apply to interconnections between settings in later life, such as the family and the peer group, the school and the world of work and, in adulthood, the family and the workplace.

Certain features consistently appear in the hypotheses regarding the developmental impact of various interconnections between settings. In regard to the isomorphism between the formal structures of micro- and mesosystems, we note further that these common features mirror the three functional parameters of the dyad: reciprocity, balance of power, and affective relation. It was stated previously that the dyad is the most versatile building block of ecological structure: it is also the functional prototype for defining optimal conditions in the operation of the mesosystem as a developmental context. Specifically, it is expected that development at this level will be enhanced to the extent that processes of interchange between settings are bidirectional, sustain and enhance mutual trust and goal consensus, and exhibit a balance of power favorable to those linking parties who facilitate action in behalf of the developing person.

The two earliest transitions that a human being typically experiences in modern societies are the temporary separation of the newborn from the mother to the hospital nursery and the move from the hospital to full time maternal care in the home. Scarr-Salapatek and Williams (1973) examined the effects of an experimental alteration in these transitions for a sample of babies born prematurely to mothers from severely deprived socioeconomic backgrounds. The authors describe the rationale for the experiment in the following terms:

Infants who are born at low birth weights to impoverished mothers are at least doubly disadvantaged. Their biological vulnerability and their subsequently poor social circumstances have been shown to interact with particularly disastrous effects upon later intellectual functioning ... A program of nursery and home stimulation was planned to demonstrate the advantage of early intervention on low-birth-weight, socially disadvantaged infants ... Scientific goals were also served in that the effects of varied stimulation in high-risk infants could be evaluated. (Pp. 94–95)

The subjects of the experiment were thirty infants weighing under eighteen hundred grams born to black mothers "from the lowest SES group in Philadelphia ... who could afford no other kind of care and who did not seek care early enough in pregnancy to enroll ... at other hospitals." An indication of the nature of the family setting and the broader ecological context in which the family lived is provided by the following description of the difficulties experienced by the investigators in conducting the second, follow-up phase of the study in the children's homes:

Maintaining contact with the mothers for over a year was difficult. Many moved every few weeks or months without forwarding addresses ...

The living conditions of ... the infants varied—some lived alone with their mothers and other relatives, some with relatives alone, and some in foster homes for all or part of the year. Many infants changed their living circumstances during the year as mothers got married, moved back with their mothers, left their mothers, and so forth ...

The mothers were typically young; only half had ever attended a prenatal clinic (Pp. 95–96)

Infants were assigned consecutively to the experimental or control group as they entered the premature nursery. In the first phase of the study, conducted in the hospital, the babies in the control group "received standard pediatric care for low-birth-weight infants. They were maintained in the isolettes and fed and changed with minimum disturbance" (p. 97). For infants in the experimental group,

the nursery staff ... were instructed before the study began to provide special visual, tactile, and kinesthetic stimulation that approximated good home conditions for normal newborns. Since standard newborn care for premature infants consists of near-isolation from patterned stimulation while in isolettes, our goal was to introduce handling, human faces and voices, and patterned visual stimulation . . .

As soon as the E infants could maintain their body temperatures for about 30 minutes (usually within 1 week after birth) they were removed from the isolettes for feeding and "play" times. The practical nurses rocked, talked to, fondled, and patted the infants during feedings in which they were held in the nursing position and could regard the nurses' faces. (P. 97)

Once babies in either group were judged mature enough, they were moved from the isolettes to open bassinets. The control infants were handled only for feeding, changing, and examinations, whereas the experimental group continued to receive special stimulation, both visual and social. Large mobiles were hung over the cribs, and "the nurses were instructed to talk to the infants, pick them up as frequently as possible when awake, and to rock and play with them around feedings" (p. 98).

As soon as the infant was discharged from the hospital, the second phase of the experimental treatment was begun: this consisted of a series of weekly visits to the home over a period of two years by a "child guidance social worker" who talked with the mother or other principal caretaker. "The visits consisted of instruction and demonstration by the social worker of stimulating child care, including observation techniques so that the mother could assess what behavior 'next steps' their infants were ready to take, and games to play which would promote 'next steps' in hand-eye coordination, reaching, grasping, vocalizing, sitting up, self-feeding, and the like" (p. 98).

No home visitors were available to mothers in the control group, although before leaving the hospital, they were provided with information on the problems and care of low-birth-weight infants and told about a "high risk clinic" that provided pediatric care through the first few years of life.

One feature of the experimental treatment having special significance in terms of a mesosystem model is that the mothers were not involved in the special program until after their children were discharged from the hospital. To be sure, this was not the investigators' original intention: "We had hoped to include the . . . mothers in the stimulation process, but this proved impractical because most were unable or unwilling to come frequently to the hospital and play with their babies. In a more advantaged group of low-birth-weight infants, the inclusion of mothers would have been practicable and important for developing a relationship between the infant and his mother in the first 2 months of life" (Scarr-Salapatek and Williams, p. 98).

Although initial measures of maternal health and neonate developmental status had favored the control group, after the stimulation program at the hospital had been in effect for four to six weeks, the experimental infants showed significantly greater weight gains and "slight to significant advantages" on the Brazelton scales. By one year, "an average difference of nearly 10 IQ points" separated the two groups. The mean score for the infants in the experimental treatment was 95, thus bringing them "to nearly normal levels of development" (p. 99), truly a remarkable achievement for a low-birth-weight sample from so deprived a socioeconomic background.

Although this important experiment does document the joint effects of experience in two different settings, hospital and home, the design does not permit a definitive assessment of the independent contributions of each, since there were no comparison groups receiving the home or hospital treatment only. Nevertheless, the research illustrates some of the parameters required of an ecological model appropriate for analyzing developmental processes for the same children in more than one setting. First, the existence of two locales (hospital and home) involves an $N + 2$ system that extends across both settings. In the present case, there are participants in four different roles. The *infant* appears in both settings, the *nurse* only at the hospital, and the *mother* and the *social worker* primarily in the home. This four-person structure permits a variety of possible subsystems and higher-order effects, both within and across settings. Unfortunately, the measures obtained focused almost exclusively on the experimental subjects—the infants—and were confined to test scores. Thus no systematic data are available about the infants' responses to the stimulation provided nor about the participants' interactions with and perceptions of each other. Here and there throughout the report are tantalizing fragments of information suggesting that certain patterns of response and relationship were central to the development processes taking place.

Newborn prematures were observed to *look* at birds suspended over their isolettes. Previously skeptical nurses (and investigators) were amazed to see 3-pound infants gazing at the brightly colored, patterned birds . . .

The infants were observed to gaze at the faces of the nurses who fed them and to respond socially to handling and voices by quieting when distressed . . .

Most mothers . . . were interested in the social worker's help, not only for their children but for themselves. They sought her advice and aid on many practical details of life . . . and in personal problems (e.g., troubles with men, mothers, siblings; feelings of depression). (Pp. 99–100)

The mothers in the experimental group were also very cooperative. Despite frequent moves, only one child was lost to the research from this group, compared with six from the control sample. Even though several of the experimental children were cared for by foster mothers for part of the year, the mothers assisted the social worker in arranging for continuation of the home visits with the new caretaker. "In no case was the home visitor excluded from an infant's home" (p. 98). Such continuity and cooperation are hardly typical in research with families from "the lowest SES group," and testify to the mothers' strong involvement with their premature infants and in the program of home visits designed to foster their children's development.

Taken together these bits of information suggest that, within the four-person system produced by the experimental treatment, certain subsystems became especially strong: nurse-infant, social worker-mother, mother-infant, and perhaps mother-infant-social worker, the last showing the second-order effect of the home visitor on the interaction of the mother with her child. Another second-order effect, in this case also temporal, may well explain the influence on the mother-infant dyad of the infants' involvement in the reciprocal relationship developed earlier with the nurses at the hospital, a relationship reminiscent of the attachment between the newborn and the mother described in the Western Reserve experiments (summarized in chapter 4).

One wonders in fact what would have happened had the mothers in the experimental group been provided with opportunities for "extended contact" of the type afforded to mothers of prematures in the study cited earlier by Klaus and his colleagues (1970). Perhaps following this experience the mothers would not have been so "unable and unwilling" to come to the hospital. What might have been the result if the researchers had made use of the apparently unexploited subsystem of nurse-social worker-mother by having the social worker begin her visits as soon as the mother returned home after delivery and report to her the nurse's enthusiastic descriptions of her premature baby's surprisingly "mature" responses? These possibilities are mentioned not as a criticism of the experiment under discussion (which constitutes a substantial scientific contribution in its present form) but as the basis for formulating our next hypothesis.

HYPOTHESIS 43
The developmental potential of a mesosystem is enhanced when the persons involved in joint activity or primary dyads in different settings form a *closed activity network,* that is, when every member of the system engages in joint activities with every other member. This pattern becomes optimal if each party interacts with every other in each setting and is subject to the qualification that the balance of power gradually shift in favor of the developing person and those primarily responsible for his well-being.

As applied to the Scarr-Salapatek and Williams experiment, this hypothesis would require that all four participants—mother, infant, nurse, and social worker—engage in some joint activity with each other both in the hospital and the home, either in dyads or $N + 2$ systems, but that gradually the balance of power shift to the mother and her infant.

Although falling short of such an ideal, the Scarr-Salapatek and Williams experiment does implement and provide support for a number of our mesosystem hypotheses regarding direct linkage. There is no evidence that either staff member accompanied mother and infant home from the hospital (hypothesis 27), but the social worker did function as a supportive link between the two settings (hypothesis 34) and engaged in joint activity with the mother when she returned home (hypothesis 36). There appear to have been no violations of boundary conditions with respect to mutual trust, goal consensus, and balance of power (hypothesis 34). Finally, the outcomes achieved are consistent with the expectation of maximum effect for young children, members of minorities, and those in weak physical condition (hypothesis 37). In terms of other types of interconnections, before going home the mothers were given information about the problems and care of prematures (hypothesis 41) and referred by the staff to a high risk clinic and were thus provided with an intermediate link to a new setting (hypothesis 38).

To be sure, the research design did not permit a test of the independent effect of each of these mesosystem measures, nor would this have been appropriate given the purpose of the study. What the experiment does provide is evidence for the effective impact of a number of these measures employed in combination.

We may note in passing that Scarr-Salapatek and Williams's findings contribute to the body of substantial evidence we have encountered in support of our basic hypotheses regarding the importance for development of the child's participation in progressively more complex patterns of reciprocal activity with someone with whom the child can develop a strong and enduring mutual attachment (hypothesis 7).

As I pointed out at the beginning of this inquiry, the overwhelming bulk of research on human development is limited to the microsystem level, consisting of studies of children in only one setting. I have been able to find only a few investigations of factors affecting the process of ecological transition, that is, the child's adaptation to a new environment. Studies of relations between settings have been even harder to discover.

Two of the researches on ecological transitions are already familiar to us. A prominent feature in Prugh and his collaborators' (1953) reorganization of hospital routines on a children's ward involved having the parents accompany the child on admission, introducing them to staff, allowing them to visit daily, and encouraging them to participate in ward care. The significant decrease in intensity and persistence of distress reactions exhibited by the experimental group constitutes evidence in support of our hypotheses regarding the salutary presence and participation in the new setting of a linking person with whom the child had previously developed a primary dyad (hypothesis 29). Schwarz and Wynn (1971), however, failed to obtain significant effects in their experimental effort to reduce the level of distress exhibited by three- and four-year-olds upon entry into day care. In a balanced design, half the children and their mothers made a twenty-minute previsit to the center a week before the child was actually enrolled; the second treatment involved having the mother remain in the setting for the first twenty minutes of the session.

Two factors may have accounted for the failure of these strategies. The first is the relatively brief duration of each treatment and the fact that it occurred only once, by contrast, for example, with the daily visits provided for in the hospital study. A second consideration is suggested by the work of Weinraub and Lewis (1977). In an observational study conducted in an experimental playroom, these investigators found that the distress exhibited by a two-year-old upon the mother's departure depended on whether and how the mother prepared the child for her impending absence: "Mothers who slipped out without saying anything had children who were least likely to play and most likely to cry; mothers who informed their children they were leaving and/or would return shortly and also gave their children explicit instructions as to what to do in their absence had children who were most likely to play and least likely to cry during maternal absence" (p. 57).

These results, in accord with hypothesis 41, emphasize the importance of providing information in advance of an ecological transition. One wonders what the results of Schwarz and Wynn's experiment would have been had the mothers who stayed been divided between those who had and those had not used the twenty-minute period to prepare the child for their departure.

To cross-validate the findings of their naturalistic investigation, Weinraub (1977) conducted an experiment in which mothers were randomly assigned to one of two conditions: one group was told to leave their children without saying anything or giving any instructions; the other were asked to explain that they were leaving, give their children explicit instructions about what to do in their absence, and reassure them they would soon be back. The results yielded support for the hypotheses, but only for boys. Once again, as in the studies of Moore (1975) and of Gunnarsson (1978), evidence suggests that male children are more likely to be affected by environmental change than females, but the phenomenon requires far more extensive and systematic documentation.

In work we have already reviewed, Pringle and Bossio (1958) reported that the retarding effects of institutionalization were not as great for children receiving visits from their families as for those who were not. A pertinent and provocative result emerges from Hayes and Grether's (1969) unorthodox analysis of achievement test scores for several thousand students enrolled in grades two through six of the New York City school system. Whereas investigators ordinarily assess academic gains by examining changes from fall to spring, Hayes and Grether also looked at the remaining interval from spring to fall—at what happened during the summer.

The results varied for children living in different circumstances. Although pupils from various social and ethnic groups started at markedly different levels in the fall and gained at somewhat different rates during the year, the main difference occurred over the summer. During the vacation, white pupils from advantaged families continued to gain at about the same rate, whereas those from disadvantaged and black families not only progressed more slowly but actually reversed direction and lost ground, so that by the time they returned to school they were considerably farther behind their classmates from more favored circumstances.[1] The authors estimate that "the differential progress made during the four summers between 2nd and 6th grades accounts for upwards of 80 percent of differences between the economically advantaged all white schools and the all Black and Puerto Rican ghetto schools" (p. 7).

The investigators state that "half or more of the differentials in reading and word knowledge are associated with non-school periods" (p. 10). On this basis they conclude that the substantial difference in academic achievement across social class and race found by the end of the sixth grade is not "attributable to what goes on *in* school, most of it comes from what goes on *out* of school" (p. 6).

Hayes and Grether also see their findings as having implications for the design of intervention programs.

If our conclusion is correct, our whole approach to equalizing educational opportunities and achievements may be misdirected. Enormous amounts of money and energies are being given to changing the school and its curriculum, retraining its teachers, and tinkering with its administrative structure—local, city, and state. We may be pouring money and energy into the one place which our results say is not primarily responsible for the . . . differentials that have been measured. (P. 10)

In light of the evidence we examined earlier for the immediate and long-range effects of day care, preschool, and other intervention efforts in a classroom setting, Hayes and Grether's statement probably goes too far, but it does point in the right direction: a key to the enhanced effectiveness of public education lies not within the school itself but in its interconnections with other settings in the society. In their investigation, the existence and importance of these interconnections—or, more precisely, their probable absence—can only be inferred from the observed effects. In another study, that of Smith (1968), the links are part of the experimental design.

This comparatively unknown investigation by an unknown researcher breaks new ground in ecological research on human development. Through an experimental intervention, Mildred Smith (1968) introduced significant changes in the prevailing relation between family and school in contemporary American society. The experiment was designed to improve the school performance of low income minority pupils in the elementary grades. The reorientation in established practice is reflected in the threefold purpose of the program, described by the author as follows: "First, it restores to the family its rightful responsibility for teaching the child. Second, it gives the family pride in being the teacher. Third, it brings together the child's 'significant others'—the parent and the teacher —as *partners,* not competitors or strangers, in the child's learning process. Neither can do this job in opposition to the other or in isolation" (p. 90).

The project involved approximately one thousand children from low income families, most of them black, attending two public elementary schools. Children of similar socioeconomic background in another elementary school in the same city were selected as controls. Among the procedures employed to achieve the stated objectives were the following.

1. To stimulate participation by parents, a group of thirty volunteer mothers were asked to assign blocks in their school district among themselves. Each then made a personal call on every family inviting parents to a planned program to "learn what they could do to help their children achieve better in school" (p. 95).

2. The program for parents was designed on the assumption that the parents wanted to help their children to succeed in school. At the initial meetings, their children's teachers explained to the parents that their help was needed. The parents were then asked to do the following: provide a quiet period in the home each day for reading and study assigned by the teacher [they were informed by the teacher that "this period is to be at a regular time so that it becomes a part of the life of the child... Remind the child of his assignment... Young children will forget" (p. 95]; listen to their children read, read regularly themselves in the presence of their children, read aloud regularly to their children, including preschool-age children, show interest in their children's work by asking questions, giving praise and encouragement when needed and deserved, prevent the school-age child's work from being damaged or destroyed by preschool children, see that the child has pencils and paper at school and at home so that he has the tools necessary for doing a good job on schoolwork, get the child to bed at a regular time each night so that he gets the proper sleep and rest, get the child up each morning with adequate time for a good breakfast, and remind the child of work papers and books that should be returned to school, since young children need this assistance. Parents who did not attend a meeting were visited by parents who had and were brought up to date.

3. A bulletin was handed each parent at the meeting. The bulletin outlined the list of objectives described above as well as related details. The contents of the bulletin were discussed in detail, and parents were encouraged to take the material home for reference. Such bulletins were provided at all subsequent meetings. Parents who had difficulty reading were given the information

orally. "Extra time and effort was undertaken to make the bulletins as neat and attractive as possible, communicating to parents that the school people had confidence in their abilities to cooperate, and —most important—respected them and considered them important. The goal was to enhance the self-concepts of parents so that they could, in the same way, enhance their children's self-concepts" (p. 95).

4. Kindergarten and primary children took a book home for their parents to read to them. On the days this was to happen, they wore tags on their lapels that read "Please read to me." Older children were given bookmarks imprinted "May I read to you?" Fathers as well as mothers were encouraged to read to their children, "thus demonstrating, particularly to boys, that men also value reading." Fathers were also invited to take turns with library duties and to serve as "male storytellers" (p. 99).

5. It was suggested that, during quiet time, parents see that preschool children were occupied, that there be no talking on the telephone (callers should be asked to call back), and that radio and television sets be turned off. (But quiet time was not to be scheduled during favorite television programs).

6. A child's dictionary was made available to each family with a child in grades four through six. "Families were encouraged to write their names on the inside cover to emphasize the satisfaction of *owning* a dictionary" (p. 98).

7. Although parents were asked to create conditions that would help the child get homework done, they were informed that "homework assignments should require no teaching by the parent" (p. 96). This meant that each parent could participate without having to be in command of school subject matter.

8. Parents were encouraged to discuss with parents of their children's friends a common time period for homework, "thus providing group support for this effort. This planning also provided that 'Johnny' would not have to leave the game when it was his turn at bat or at the time when it was his turn to shoot at the 'purey'" (p. 97).

9. Teachers agreed to limit home assignments to fifteen minutes for children in the primary grades and thirty minutes for those at upper elementary levels. Every morning each child reported whether the assignment had been done. "Thus the child was checked on whether or not he had completed the task rather than how well it was completed. Every child could, therefore, be successful provided that his parents were giving the needed support

at home. If a child frequently failed to complete a task, the parent was sought for a conference ... This record keeping by teachers assured each child that the two people concerned with his academic achievement were in constant communication with each other" (p. 97).

10. Teachers were provided with clerical assistants. Business students typed and duplicated materials prepared by the teachers and provided other services "thus freeing teachers to give more personal attention to students. This was one of the greatest morale boosters for teachers that the program provided" (p. 102).

11. An inservice program for teachers emphasized the influence of environmental factors on children's classroom behavior and performance. Teachers were helped to understand "that the problem of the underachiever is not necessarily one of insufficient capacity or ability, but is frequently one of inadequate environmental support or motivation" (p. 93).

12. "An educational director from a local factory visited the schools and showed slides of people performing various skilled jobs. He explained the advance high school courses required for eligibility in his training programs and reminded the children that competence in science, reading, spelling, and mathematics would make them eligible for courses" (p. 101). In addition, blacks in the area who held skilled jobs visited classrooms, explained work, and "told how their elementary school subjects had been important to them later in their lives" (p. 102).

Smith's transforming experiment involves virtually all the interconnections that have been stipulated for a mesosystem model as well as for the hypotheses defining the mesosystem factors that influence the capacity of settings to enhance development. Processes of interchange of every variety are set in motion not only between the school and the home but also between the school and the world of work, along with the local neighborhood. Across these settings, opportunities are provided for establishing a variety of links and transcontextual dyads that call for joint activity and facilitate the development of primary relationships. There are $N + 2$ systems and second-order effects across every border. The circle of significant adults and settings in the child's life is extended, and each member engages in joint activities both with the child and with each other in every major setting. Virtually all modes of intersetting communication are employed, and considerable information is offered in each setting about the others. Although no provision is made for facilitating ecological transitions into or within the school, attention is paid to preparing students for ultimate entry into the world of work. With respect to the overall relation among settings, major emphasis is given to reciprocal interaction between settings, overlap in activities, the building of positive attitudes in both directions, and the insuring of complementarity of roles and a balance of power such that both parents and teachers can exercise and maintain a sense of control in their respective domains of activity. Perhaps the only question to be raised about the program is whether pupils are given adequate opportunity to move toward greater self-direction in the course of their school experience.

On the side of the dependent variables, however, the study has some serious shortcomings; the imagination and comprehensiveness demonstrated in the planning of the experimental treatments were not matched in the selection of outcome measures. Results in quantitative form are limited to significant gains on tests of reading achievement and in parents' reactions to the program solicited in a questionnaire. The questionnaire brought a gratifying 90 percent return. Particularly favorable attitudes were expressed toward the home study program and reading experiences: that and the test results are all we know about the accomplishments of the experiment. Had the research included at least some data on the roles, activities, and relations exhibited by experimental and control subjects in the classroom setting and, perhaps even more significant, in the family and in the neighborhood peer group, the contribution of the study to both science and educational practice would have been immeasurably enhanced.

Neverthless, the Smith study offers a prototype for research designs and experiments in the ecology of human development. Furthermore, it confronts a pervasive problem in contemporary American society.

As I have indicated elsewhere (Bronfenbrenner, 1970a, 1974b, 1974c), the school is becoming increasingly isolated from the home. As neighborhood schools disappear, the school buildings become farther away, larger, and more impersonal. The staff increase in number, are drawn from a larger area, and often commute rather than live in the local community. As a result parents and teachers are less likely to know each other at all.

Moreover, as schools are moved to the outskirts of town, they become compounds physically and socially insulated from the life of the community, neighborhood, and families the schools purport to serve as well as from the life for which they are supposedly preparing the children. The insularity is repeated within the school itself, where children are segregated in classrooms that often change yearly. Moreover, the classrooms have little or no social identity of their own and little connection with each other or with the school as an active community. This comparative absence of communal life gives freer rein to the disruptive forces of age segregation as large numbers of children are thrown together primarily with their peers. It is significant that the only adults who are encouraged to enter this children's world tend to be persons with master's degrees, often having backgrounds that do not reflect a rich diversity in experience of the world.

As a result of these trends, the school has become over the past two decades what I have called "one of the most potent breeding grounds of alienation in American society" (Bronfenbrenner, 1974b, p. 60). In my view it is this alienation that underlies the progressive decline observed in achievement test scores that has been recorded over the past dozen years both for the college-bound and for the general population of students at the elementary and secondary levels (Harnishfeger and Wiley, 1975). Its more acute manifestations are seen in rising rates of homicide, suicide, drug use, and delinquency for children of school age (Bronfenbrenner, 1975).

One barometer of these destructive trends is the rising level of vandalism and violence in the schools. A report of the Committee of the Judiciary of the United States Senate telegraphs its major findings in its title: *Our nation's schools—a report card: "A" in school violence and vandalism.* The report emphasizes that the pattern is not restricted to big cities and their slum areas but is now a national phenomenon. No school is without a security budget and often a security force. Literally, the handwriting is on the wall. Although this judgment may represent a premature conclusion based on inadequate data about a complex social phenomenon, as a hypothesis for stimulating and guiding rigorous empirical work it has an important scientific function. It is deliberately included to suggest the kind of twoway interaction between developmental research and public policy that I believe to be essential to advance basic knowledge of the forces that shape the course of human development.

From the perspective of our theoretical model, the alienation of children and youth and its destructive developmental sequelae are mesosystem phenomena. They reflect a breakdown of the interconnections between the various segments of the child's life— family, school, peer group, neighborhood, and the beckoning, or all too often indifferent or rejecting, world of work. It therefore becomes the social responsibility, as well as being an unparalleled scientific opportunity, for the researcher on human development to undertake field and experimental studies that will illuminate the nature, consequences, and potential of these interconnections.

If it is true that ecological transitions and interconnections between settings play a major role in affecting the direction and rate of development, then adulthood should be a period of dramatic shifts, spurts, and slumps in psychological growth. For although biological changes occur at a slower rate after adolescence, social changes in contemporary industrialized cultures proceed apace. The young person leaves home to enter a variety of settings as he goes to college, finds a job, joins organizations, gets married, becomes active in community life, changes jobs, and so on. In some instances the old settings are not abandoned for the new but continue to be frequented so that the network expands with ever greater possibilities of interconnection. The most stable and enduring base throughout this process, current divorce rates notwithstanding, remains the family.

Our theory predicts that the variations among individuals and groups in the number and nature of these transitions and intersetting relations should produce differences in development reflected by diversity in the range and level of molar activities, role repertoires, and patterns of social interaction at successive stages of adult life. The data needed to verify these expectations, however, are exceedingly sparse. As we have noted, research on human development has until now been concentrated at the two extremes of childhood and old age, with more emphasis on the former. The reason, I suggest, is the absence of a coherent theoretical framework for conceptualizing sources, processes, and outcomes of development in periods of life not characterized by rapid and readily detectable biologically mediated changes. It is my hope that the ecological schema developed here can provide a base for a much-needed expansion of systematic research on human development in the middle years of life.

The best data I have been able to find concerning mesosystem influences on adult development emerge, paradoxically, as offshoots of studies on the developmental impact of settings in early childhood. We saw earlier that mothers who placed their children into day care or preschool programs were themselves affected by the changes in their child. These maternal effects were thus manifested outside the day care center, and even outside the home, and hence represented mesosytem phenomena.

Dramatic effects on the mother's development were also observed in conjunction with home-based programs involving both mother and child. Witness the following account by Karnes and her associates:

The competence and capabilities demonstrated by the mothers within the program were reflected in increased community involvement. Four mothers assumed responsibility in the summer recruitment of Head Start children, and one was hired as an assistant teacher and promoted later to the position of head teacher. Two mothers spoke of their experiences in the mother training program at a Head Start parent meeting. Finally, total group involvement was demonstrated at a local Economic Opportunity Council meeting called to discuss the possibility of establishing a parent-child center in the community. Twelve of the 15 mothers attended this meeting and were, in fact, the only persons indigenous to the neighborhood in attendance. (1970, pp. 931–932)

A similar effect is reported by Gilmer and her coworkers for a program involving both home-based and preschool intervention.

Not reported in the results section is a careful study that was made of the changes in life style of the mothers in the treatment groups . . . To the extent that one may attribute the life style changes to the involvement of the mothers in the program, we have here some of the most interesting results of the study. These findings, however, should certainly be interpreted with caution because, over a period of two and one half years in the late 1960's, many social changes were taking place.

Still we find that many of the mothers went on to finish their high school education and enrolled in training courses to upgrade vocational skills. Several have taken positions in preschool and day care centers. Five of the mothers at one time were functioning as home visiting teachers themselves.

Interest and participation in community affairs broadened. Social contacts with other members of the community increased markedly. There were cooperative outings, a rotating book library, and the establishment of a bowling league which included fathers. One somewhat ironic effect of the program, from the standpoint of maintaining statistical control, was the wish of many of the parents to move out of the housing project to more improved housing. There were increases in the number of checking and savings accounts, which almost none of the parents had before the study began.

These changes in life style would seem to be the result of the development of environmental mastery, which may be expected to have a supporting effect on the children's continued development. (1970, pp. 47–48)

It is clear that the participation of either the mother or the visitor as links between the home and the preschool program affected not only the intellectual performance of the child (all the children showed significant gains in IQ) but also the mother's own development as reflected in an increased level and range of molar activities outside the home.

In a study of development in old age, Aldrich and Mendkoff (1963) exploited an experiment of nature provided by the impending closing of a home for disabled elderly persons and the transfer of residents to other similar homes. The process of relocation lasted approximately two years and involved 233 patients, 70 percent of whom were aged seventy or older. Length of residence in the home prior to transfer ranged from one to forty-five years. "Patients ... were relocated from one institution to another without regard to their state of health or family relationships but solely as the result of administrative necessity ... Transfer was made primarily to nursing homes of substantially the same or better quality in the same community" (pp. 185–186).

Of interest to the investigators was the general issue of the impact of relocation, actual and anticipated, on the well-being of the residents. The principal outcome measures were deviations from expected mortality rates for age during three successive periods: before patients learned of the projected relocation, while patients were still in the home while awaiting transfer, and after relocation had taken place. Age-adjusted expectancies were calculated on the basis of observed mortality rates in the home for the preceding decade. In addition, the investigators "evaluated the response pattern of each patient who was sufficiently aware of the news of relocation to respond in a discernible manner" (p. 404). Reactions were classified under such headings as the following: philosophical, angry, depression, and denial.

The analysis of mortality as a function of months elapsed since relocation revealed that the number of deaths in the first three months was three and a half times higher than during the remainder of the year, when the observed rate again fell to the expected level. Mortality rates were also higher than projected during the period of anticipated relocation, but this effect was not significant. Finally, variations in death rate were examined as a function of the patient's reaction to the news of the impending move. The following reliable interaction effects were found: "The survival rate was highest for patients who took the change in their stride or were overtly angry;

patients who became anxious but did not withdraw, survived reasonably well; and patients who regressed, became depressed or denied that the Home was closing, survived less well" (p. 190).

Actual mortality rates ranged from under 5 percent for the least vulnerable group to over 45 percent in those who reacted with depression or denial. The patients who were most likely to survive were those who acknowledged and coped with the coming crisis either by anger or explicit acceptance. Least likely to survive were those who retreated from the conflict situation by depression or denial.

In light of their findings, the authors offer the following recommendations:

Since patients either die or become adapted within three months, efforts to assist adaptation to relocation should be concentrated within this period . . . Subjectively, casework was apparently helpful in the patient's preparation for relocation and probably was helpful in reducing mortality . . . Ideally, for control purposes, we should have limited the casework help to alternate residents of the Home; practical and humane considerations, however, did not permit casework to be withheld from any group of patients. Therefore, it was not possible to determine whether the death rate for any subgroup would have been materially higher without such preparation.

Despite conscientious efforts, the dangerous effects of relocation cannot be completely eliminated. The best prevention is not to relocate elderly and disabled people. This type of prevention is not always, possible, however. The danger remains greatest among helpless and psychotic elderly people who, on the basis of practical considerations, are most likely to require relocation in other institutions. (Pp. 192–193)

The results of this study lend further support to several of our hypotheses regarding the mesosystem properties most conducive to sustaining development. They underscore the special vulnerability of particular groups, in this instance the elderly disabled, to isolated settings (hypothesis 37) and their heightened need for information, advice, and preparatory experience prior to a transition to a new setting (hypothesis 41). It would have been instructive to discover in addition whether, in accord with hypothesis 27, mortality rates were lower among patients who were transferred to the new institution in the company of their friends (or, in line with hypotheses 35 and 36, moved into a home at which one or more friends were previously or subsequently relocated), who were more frequently visited by family or friends both in the old and new location, or who were transferred to the one facility that "had accepted 28 patients from the Home . . . as well as many employees and some of the activities various religious groups had sponsored" (p. 186).

What are the ecological transitions and intersetting connections that are most important to investigate in terms of their impact on developmental processes? For development in childhood and adolescence, the available evidence appears to point to a trio of settings involving home, school (including day care center and preschool), and peer group. For adult development, the paucity of data makes the question more difficult to answer, but some clear indications emerge from a pilot study conducted in the first stage of a cross-cultural project on family support systems (Bronfenbrenner and Cochran, 1976; Cochran and Bronfenbrenner, 1978). The pilot study was designed to pretest an instrument for assessing sources of stress and support experienced by parents of young children. In a sample of seventy families, the most pervasive source of stress was found to be conditions at work, in particular, working hours and required overtime on evenings and weekends. Conversely, in those instances in which it was present, flexibility in job schedule was identified by parents as a source of support second only to the availability of satisfactory child care arrangements. Moreover, these respondents identified their work as one of the main avenues beyond their parental role for achieving personal fulfillment, the women more so than the men.

In sum, in a pilot study taking the home and family as its point of departure, the world of work emerged as a key setting of the mesosystem in adulthood. Job conditions were perceived as potent forces affecting the respondent's ability to function as a parent, and presumably such influence affects the child. Yet in American society at least, the parent's workplace is not one that children enter often or for very long periods of time. Its status in an ecological schema oriented around the child is therefore that of an exosystem domain. As we shall see, it is peripheral only in its position, not in its power to determine possibilities and processes in the child's development.

10.

The Exosystem
and Human Development

An exosystem has been defined as consisting of one or more settings that do not involve the developing person as an active participant but in which events occur that affect, or are affected by, what happens in that setting. It follows that to demonstrate the operation of the exosystem as a context influencing development it is necessary to establish a causal sequence involving at least two steps: the first connecting events in the external setting to processes occurring in the developing person's microsystem and the second linking the microsystem processes to developmental changes in a person within that setting. The causal sequence may also run in the opposite direction. The developing person may set in motion processes within the microsystem that have their reverberations in distant quarters. In either case, it must be shown that a two-stage sequence has occurred.

Very few existing studies meet this double requirement: instead, one or the other link is usually taken for granted. Thus there are two common patterns of research on the developmental effect of environmental influences outside the immediate setting containing the child. In one, the investigator demonstrates the impact of external forces on processes occurring within the setting and assumes, or leaves it for others to assume, that these processes have developmental consequences. The other strategy leapfrogs the intermediate phase by demonstrating a connection—more often simply a statistical association—between some aspect of the larger external environment and some developmental outcome, bypassing any microsystem processes that might have been involved.

So frequent is the resort to one or the other of these abridged strategies, and so rare the explicit demonstration of the two-step sequence, that I am left little choice but to cite examples of pre-

237

sumed interconnections that are incomplete, that have one or another part missing or present only by implication. Nevertheless, I have tried to select illustrations that approach the ideal, or that at least acknowledge what links are not established in the research design.

The failure to make the final connection in a causal sequence beginning in an exosystem occurs most frequently in studies of external influences affecting socialization within the family. Here the forgotten figure is the child. Convincing evidence is presented for differences in patterns of parent-child interaction, and the assumption is made, often not unreasonably, that these differences will affect the child's behavior and development. The first such research cited here is particularly instructive for, without presenting any data on the actual behavior of children, it demonstrates that the child can influence the actions of parents no less than they can affect him. Moreover the child's influence reaches out beyond the family into settings that he never enters—that are a part of the child's exosystem.

The only investigation I have found that systematically examines the relation between parental social networks and social interaction within the family bypasses the issue of a two-step process by focusing only on possibilities for direct connection between the parties involved. McAllister and his colleagues (1973) examined the social interactions of parents both within and outside the family as a function of the presence of a retarded child in the home. Unfortunately the study design exhibits a characteristic limitation of the traditional research model: systematic behavioral data are provided only for the subjects of the study—in this instance, parents. The investigators employed a homogeneous sample of Anglo-American families with children living in a southern California city. Of the total number of 1065 households, 281 had one or more "behaviorally retarded children" as defined by scores on a specially constructed test adapted from the Vineland (Doll, 1953) and Gesell (1948) developmental scales. Intrafamily interaction was measured from the mother's report of the frequency with which parents read stories to their children or talked with them about "their friends, problems, and things like that" (p. 96). Extrafamily interaction was assessed by parental membership in voluntary organizations and contacts with relatives, neighbors, friends, and coworkers, again as reported by the mother.

In line with the researchers' expectations, families with retarded children showed lower levels of parent-child interaction within the home. As the authors properly point out, "The data do not allow a determination of cause-and-effect and so, the question remains whether there is less interaction in the family because of behavioral retardation or whether the child is retarded as a result of low family interaction" (p. 97). The results regarding extrafamilial interaction differed on the dimension of formality. There was no relation between the presence or absence of a retarded child and parental membership in organizations, but there was a reliable tendency for parents of retarded children to visit less frequently with neighbors and, to a lesser extent, with relatives. This trend was especially marked for mothers. In addition, fathers with a retarded child were significantly less likely to visit with coworkers. Interaction with friends showed a similar trend for both parents, but the difference was not reliable.

The authors interpret this pattern of results as reflecting a tendency for parents with retarded children to disengage themselves from activities and relationships outside the family as a function of the prominence of the child in the given social context. Thus differential participation was most pronounced in the neighborhood, "where the retarded child is most visible." The area least affected was membership in formal organizations, since the parent "can participate without the necessity of it becoming known that the family has a retarded child." The marginal magnitude of the observed relationships for relatives and friends was seen as the product of a somewhat different dynamic: since "in intimacy . . . impairment cannot be hidden" (p. 98), it does not exert as marked an influence on social activity within these closer spheres.

The failure to find differences in the sheer number of organizations joined by parents with and without a retarded child prompted the investigators to propose a hypothesis differentiating organizations by type.

We . . . suggest that those families of the retarded who belong to impairment-oriented organizations would be more likely than those who belong to traditionally oriented organizations to participate with friends, neighbors, and co-workers on a normal basis. This hypothesis is based on the assumption that those parents who participate in impairment-oriented organizations have come to terms with the retardation of their child, whereas those who have not come to terms with the retardation are more likely to "withdraw" and participate less with neighbors, friends, and co-workers as well. (P. 98)

For reasons that they do not specify, the authors state that they were unable to test this hypothesis. The failure is all the more regrettable in that the formulation represents an exosystem analogue to hypothesis 34, which emphasizes, at the level of the mesosystem, the importance of goal consensus between settings and of roles compatible with action in behalf of the developing persons. The exosystem version is different only in the property that differentiates the two system levels. In a mesosystem, the developing person is active in *both* settings along with the supplementary link. In the present study, if we consider the developing person to be the retarded or nonretarded child, that person is located primarily in the home. Since it was the parent's social interactions in the community that were assessed, they provided the link, examined in this study, between the family and the extrafamilial settings that constituted an exosystem for the child.

Neither hypothesis 34 nor any other hypothesis specifying the optimal structure of the mesosystem (35 through 42) requires that the developing person be present in all the settings under consideration. Thus these hypotheses, as stated, can be applied either to the meso- or to the exosystem. The hypotheses are in fact valid at both levels. In other words, the forms of linkage, communication, and availability of knowledge that define the optimal properties of a mesosystem from the perspective of human development constitute the optimal conditions for exosystems as well.

I lamented the inability of McAllister and his colleagues to test their exosystem hypothesis about parents' social participation outside the community. Even more regrettable, from our perspective, is the absence of data that would permit examining the relationship between intra- and extrafamilial interaction, on the one hand, and the behavior and development of the children, both retarded and nonretarded, on the other. Were such information available, it would be possible to test the principle embodied in hypothesis 35 at an exosystem level. The principle affirms that the developmental potential of a setting is enhanced by the existence of supportive links with external settings. Applied to the families in the McAllister study, this proposition leads to the prediction that parents' interaction with their children will have greater impact the more numerous the supportive links between the family setting and the external environment. Translated into statistical form, this prediction implies a higher correlation between measures of parent-child interaction and developmental outcomes for families with extensive social networks than for those with few links outside the home. Moreover to the extent that, following the investigators' hypothesis, external contacts may involve negative experiences for parents with a retarded child, the posited differential in the magnitude of correlations should be smaller for such families than for those with normal children. The predicted patterns would be analogous to the contrasting results obtained by Klaus and his colleagues (1970) (who also provided no data on the child's behavior) when opportunity for extended contact was provided to mothers of normal and premature newborns: the experimental effect turned out to be more pronounced with the former than the latter. Here again is an illustration of the isomorphism between relationships at different levels of the ecological environment, in this instance an exosystem involving parental social networks and the microsystem of a hospital delivery room.

The tendency to report features of the child's microsystem but stop short of describing the child herself is even more common in studies that select as independent variables conditions or events occurring in the child's exosystem. Again I have chosen examples that approach a full exosystem model to the extent that the assumption of developmental effects appears quite justifiable.

Research on environmental factors conducive to child abuse, or its prevention, is a case in point. In a study of child neglect among low income families, Giovanoni and Billingsley (1970) sought to identify the environmental conditions associated with the parents' treatment of the child. Predictive of the child's mistreatment were such factors as number of children, single parenthood, inadequate housing and sleeping arrangements, absence of a telephone or wrist watch, and other correlates and consequences of extreme and prolonged poverty status. Two environmental factors in the lives of these families had a preventive effect: the existence of a functional kinship network and church attendance. In evaluating their findings, the authors conclude that "Among low-income people, neglect would seem to be a social problem that is as much a manifestation of social and community conditions as it is of any individual parent's pathology" (p. 204).

Corroborative data come from a large-scale correlational analysis of child abuse reports and socioeconomic and demographic information for the fifty-eight counties of New York State (Garbarino, 1976). In the investigator's words, "A substantial proportion of the variance in rates of child abuse/maltreatment among New York State counties (three samples) was found to be associated with the degree to which mothers do not possess adequate support systems for parenting and are subjected to economic stress" (p. 185).

Most research on the role of television in the lives of children has focused on its direct effects, primarily in the arousal of aggression and violence (Liebert, Neale, and Davidson, 1973). An ecological analysis suggests the possibility of another, more roundabout process no less disturbing in its potential outcomes. As I have written elsewhere, "Like the sorcerer of old, the television set casts its magic spell, freezing speech and action and turning the living into silent statues so long as the enchantment lasts. The primary danger of the television screen lies not so much in the behavior it produces as the behavior it prevents—the talks, the games, the family festivities and arguments through which much of the child's learning takes place and his character is formed" (Bronfenbrenner, 1974c, p. 170).

In search of the research literature bearing on this issue, Garbarino (1975) was able to identify only one investigation that dealt with the question directly. In a field survey Maccoby (1951) found that 78 percent of the respondents indicated no conversation occurring during viewing except at specified times such as commercials, and 60 percent reported that no activity was engaged in while viewing. On the basis of her findings, Maccoby concluded, "The television atmosphere in most households is one of quiet absorption on the part of family members who are present. The nature of the family social life during a program could be described as 'parallel' rather than interactive, and the set does seem quite clearly to dominate family life when it is on" (p. 428).

Maccoby's study was published over a quarter of a century ago and no further research on the problem has been undertaken since that time. With the rapid growth of television and the television culture in the intervening years, the impact of the medium on family life has become both more pervasive and more profound. The question of how any resultant change in family patterns has in turn affected the behavior and development of children remains completely unexplored.

Since the television program enters the home from an external source, it constitutes part of the child's exosystem. To the extent that this powerful medium exerts its influence not directly but through its effect on the parents and their interaction with their children, it represents another instance of a second-order effect, in this case operating not completely within a microsystem but rather across ecological borders as an exosystem phenomenon. Thus once again we see the isomorphism of relationships at the different levels of ecological structure.

Another abridged strategy in research on exosystem influences is to document differential outcomes in the behavior of the child or other developing person but to overlook—conceptually, operationally, or both ways—the intervening contexts and processes that link the external conditions or events to the observed developmental change. Often the connection is assumed to be immediate and direct without consideration of more complex but equally plausible processes.

A pertinent example is provided by an elegant ecological study of the influence of apartment noise on auditory discrimination and reading ability in children (Cohen, Glass, and Singer, 1973). The following summary of the design and the data constitutes the author's abstract:

This study examined the relationship between a child's auditory and verbal skills and the noisiness of his home. Expressway traffic was the principal source of noise. Initial decibel measurements in a high-rise housing development permitted use of floor level as an index of noise intensity in the apartments. Children living on the lower floors of 32-story buildings showed greater impairment of auditory discrimination and reading achievement than children living in higher-floor apartments. Auditory discrimination appeared to mediate an association between noise and reading deficits, and length of residence in the building affected the magnitude of the correlation between noise and auditory discrimination. Additional analyses rule out explanations of the auditory discrimination effects in terms of social class variables and physiological damage. Partialling out social class did, however, somewhat reduce the magnitude of the relationship between noise and reading deficits. Results were interpreted as documenting the existence of long-term behavioral after-effects in spite of noise adaptation. (P. 407)

The investigators viewed their research as a real-life counterpart to laboratory experiments demonstrating degradation of task performance as a direct after-effect of exposure to noise, and interpret their results in the same terms. The two situations are not precisely analogous, however, since the real-life setting included other persons besides the children who were selected as the subjects of the study. Moreover, these other persons, who were the child's parents and other members of the family, were also exposed to traffic noise and in all likelihood affected by it. The possibility remains that the impairment of the child's auditory discrimination and verbal skills came about not only as a function of his own difficulties in hearing, and sustaining attention, in a noisy environment, but also because others around him, notably his parents, were similarly affected and engaged him less frequently in conversations, reading aloud, or correction of the child's verbal utterances. No data are available to demonstrate or disconfirm the existence of such a second-order effect, but relevant information could readily have been obtained had the other participants in the setting been included in the research design and interviewed about the nature and frequency of activities involving verbal interaction with or in the presence of the child.

It is possible to give full recognition to the family as the mediating agent between the external environment and the child and yet fail to demonstrate that any mediating process has actually taken place. To consider a common example, innumerable studies have been published documenting social class differences in the behavior of children (for a comprehensive summary, see Clausen, 1966; Hess, 1970). In most of this work, differential patterns of socialization within the family are identified as the immediate source of the observed effects, but the causal connection is not explicitly established. To be sure, an equally large number of researches show just as impressive differences by socioeconomic status in parents' child-rearing values and practices (see reviews above); implications for development are persuasively drawn but not actually demonstrated. In these respects research on social class and socialization represents a prototype of the kind of investigation into exosystem influences on development in which either the child or the family is left out of the empirical equation.

What of the less frequent but still substantial number of studies (see reviews above) in which data on differences by socioeconomic status are provided for parents and children in the same family? Are these not examples of full-fledged exosystem analyses in which both steps in the causal sequence are demonstrated rather than assumed? Unfortunately, there are still two missing pieces to the puzzle. The first is methodological. The fact that families from different social classes exhibit consistent differences both in parental socialization practices and in measures of developmental outcomes for the children does not establish that the latter are a function of the former, for both could be direct products of other variables associated with or actually defining socioeconomic status, such as income, education, occupation, family size, single parenthood, and so on. To provide persuasive evidence of causation, it would be necessary to demonstrate at the very least that patterns of child rearing and measures of developmental outcome are significantly correlated *within* as well as across class, that is, after controlling for such confounding factors as parental income, education, occupation, marital status, and family size. To my surprise, I have not been able to find any investigation in which such analyses were carried out.

There is a second problem with the existing studies in this area. It is substantive rather than methodological and, from an ecological perspective, constitutes an even greater shortcoming. In developmental research, if not in social science in general, social class has typically been treated as a linear variable rather than analyzed in systems terms as an ecological context. To do so would require examining the settings that are implicated in the operational definitions of socioeconomic status and the roles, activities, and relations in which persons entering these settings necessarily become engaged.

Since social class is usually defined in terms of income, occupation, education, and occasionally place of residence, the settings involved are the workplace, the school, and sometimes the neighborhood. If one is interested in the developmental consequences of socioeconomic status, there are two questions that should be asked. First, how do these settings differ in the roles, activities, and relations that they require of persons living in diverse socioeconomic strata? Second, what are the effects of this differential experience on the development of these persons?

I know of only one scholar who has recognized the scientific importance of these questions and pursued the answers steadfastly in his empirical work. Understandably he has not attempted to address all three of the setting domains implicated in definitions of social class but has wisely concentrated on only one of them—parental occupation—although his work contains important findings for the school as well. In a series of studies conducted over the past two decades, Kohn and his colleagues (1963, 1969, 1977; Kohn and Schooler, 1973, 1978) have explored systematically the role requirements of the job and their impact on the development of the jobholder not only at work, but in the other major setting in life—that of home and family. Although departing radically from established social science approaches in his treatment of social class, Kohn—unfortunately from our viewpoint—adheres to the first of the two prevailing strategies for studying exosystem influences: looking at effects on parents but stopping short of the effects on children themselves.

The general research question Kohn addresses is by no means a simple one. Noting that socioeconomic status is a pervasive source of variance in virtually every aspect of human activity, he seeks to find out "what is it about 'class' that makes it important for so much human behavior" (1969, p. 3). Operationally, Kohn starts with a more restricted focus: the relationship of social class to the development of values. What is significant about his work, however, is not where he begins but where he ends. Without explicitly employing an ecological model, he commences with what is clearly a macrosystem phenomenon, the class structure of the society, and then traces its manifestations at a microsystem level in specific life settings. Working with samples of adult married men with children, Kohn finds that the principal difference in value systems associated with social class position focuses around the issue of self-direction versus conformity to external authority, whether in relation to parental values, self-conceptions, attitudes toward work, or orientation to society.

The higher a person's social class position, the greater the likelihood that he will value self-direction, both for his children and himself, and that his orientational system will be predicated on the belief that self-direction is both possible and efficacious. The lower a person's social class position, the greater the likelihood that he will value conformity to external authority and that he will believe in following the dictates of authority as the wisest, perhaps the only feasible, course of action (1977, p. xxvi)

These values in turn affect the adult's behavior, specifically in the realm of parental practices, as assessed from independent reports by the parents and their children. "The evidence ... clearly indicates that middle class parents' higher valuation of self-direction, and the working class parents' higher valuation of conformity to external authority, influence their disciplinary practices and also the allocation, between mother and father, of responsibility for providing support to, and imposing constraints on, their children" (p. xxxiii).

Having established these differential patterns, Kohn proceeded to seek out the particular features of a person's class position that accounted for their emergence. His first step was to evaluate the possible contribution of a host of psychological and social factors associated with socioeconomic status. The results were essentially negative.

The relationship of class to parental values is not a function of parental aspirations, family structure, or—insofar as we have been able to measure them—family dynamics. The relationships of class to values and orientation in general are clearly not a function of such class-correlated dimensions of social structure as race, religion, or national background. Nor are they to be explained in terms of such facets of stratification as income and subjective class identification, or of conditions that impinge on only part of the class hierarchy, or of class origins or social mobility. Finally, the class relationships do not stem from such important (but from our point of view, tangential) aspects of occupation as the bureaucratic or entrepreneurial setting of jobs, time-pressure, job-dissatisfaction, or a host of other variables. Any of these might be important for explaining the relationship of class to other social phenomena; none of them is important for explaining why class is consistently related to values and orientation (1969, pp. 189–190)

He then examined the variables included in the measure of social class itself, which is typically based on some combination of occupation, education, and income. Kohn therefore sought to identify the relative influence of each on the value orientation of his adult male subjects. Income (along with the person's subjective identification of his own class position) turned out to be least important. The key factors were education and occupational position, with the former predominating: "Education is the more potent of the two dimensions, being more strongly related to parental values, to self-values, to judgments about the extrinsic feature of jobs, and—most strongly of all—to authoritarian conservatism" (p. 132).[1]

Moreover, the effects of education apparently exerted their influence independent of job status. "The relationship of education to values and orientation is not greatly affected by occupational experience or by any other experiences that we have examined. The importance of education for men's values and orientation—at least under the conditions of life in the United States in the mid-1960's—is great, no matter what conditions men subsequently encounter" (p. 191).

What was it about education that produced these effects? Although as reported below Kohn undertook a systematic analysis of the work situation, he did not do so for schooling. Information in this sphere was confined to the number of years of education, treated as a linear variable.[2] Kohn does offer a relevant hypothesis and presents some indirect evidence for its support: "Education provides intellectual flexibility and breadth of perspective that are essential for self-directed values and orientation; lack of education must seriously interfere with men's ability to be self-directed" (p. 186). Consistent with this interpretation, statistical control on a measure of flexibility (based on how well men dealt with problems in the interview) substantially reduced the correlations between educational level and value orientations.

Paradoxically, an indication of which aspects of the school experience foster intellectual flexibility is provided by the results of Kohn's analysis of the principal setting he chose to explore—the father's work situation. On the basis of interview data, each job was analyzed in terms of what ultimately became three major parameters: the substantive complexity of work in the realm of things, data, ideas, and people; the degree of routinization and repetitiveness in required tasks; and the closeness of supervision. These aspects of the job situation were then related to each man's value orientations in the areas of parenthood, work, the self, and society in general (including such parameters as authoritarian conservatism, criteria of morality, trustfulness, and stance toward change). All these spheres were affected by certain critical aspects of the work experience. Kohn found that "the crucial occupational conditions are those that determine how self-directed one can be in one's work—namely, freedom from close supervision, substantively complex work, and a nonroutinized flow of work. These occupational conditions are empirically tied to valuing self-direction and to having an orientation to oneself and to the outside world consonant with this value" (1977, p. xxxiv).

Elaborating upon an idea originally suggested by Bowles and Gintis (1976), I propose that a parallel analysis of educational experience, particularly as a child moves from the elementary grades, through junior and senior high school, into college, would reveal differences along these same parameters that in turn would be correlated with changes in value orientations and corresponding behavior in all areas of life. The further one goes in school, the more likely one would be to experience freedom from close supervision, nonroutinized flow, substantively complex work, and opportunity for self-direction. If level of schooling is held constant, it is again along these parameters that the experiences of lower versus upper class children would be most apt to differ both in school and out. This line of thinking argues for the importance in educational research of investigating systematically the changes in activity that occur from one grade to the next, from one school to another, and for pupils from different socioeconomic, ethnic, and cultural groups within a given educational setting.

Although Kohn employs a different terminology, it is clear that his three critical features of the work situation represent, in terms of our ecological schema, specifications of particular kinds of activities, roles, and relations affecting adult behavior and development in the job setting. In addition, they closely correspond with our hypotheses, derived from research conducted with adults and children in quite different settings, regarding the importance for psychological growth of engagement in complex molar activities in a social context characterized by a balance of power favorable to the developing person.

Finally, by demonstrating that the orientations generated in the work situation carry over into the home and are reflected in the adult's values and practices as a parent, Kohn's findings point to the powerful influence of the work setting as an exosystem indirectly affecting the development of the child. It is significant in this regard that the family's socioeconomic status, based on the father's occupational level, predicted the mother's parental behavior better than the father's (1969). The reason, in Kohn's view, is that children's behavior is less likely to be the father's concern. From an ecological perspective, the fact that the father's situation at work affects the mother's treatment of the child signifies the operation of a three-person structure across borders within an exosystem.

As with the other researches we have reviewed, it remains to be demonstrated that the particular patterns of parent-child interaction set in motion by values and behavior orientations generalized in the father's workplace do in fact produce developmental changes in the child as reflected in his activities when alone or with other persons in different settings.

That in all likelihood such changes will be found once research designs are employed that permit the demonstration of the two-stage causal sequence is indicated by a post hoc analysis of interview data from a transforming experiment conducted in Mexico and from the results of a field study of development of minority group children in a California city.

Working in poor residential areas in Mexico City, Almeida (1976) offered an eight-week training course in child development, in one case for teachers alone, in another for teachers and parents together. In each of six neighborhoods, one sixth grade classroom was randomly assigned to the experimental treatment (parents plus teacher) and another to the control group (teachers only). The weekly two-hour training sessions were conducted by persons who lived and worked in the immediate neighborhood. The investigator hypothesized that parental participation would result in enhanced motivation and learning on the part of pupils as a function of increased mutual understanding and convergent value commitments on the part of parents, teachers, and children.

Almeida's findings are instructive both methodologically and substantively. The difference between the experimental and control group turned out to be significant on most outcome measures when tested against individuals within treatments, as is typically done in psychological experiments. But none of the treatment effects was significant when tested against an appropriate error term based on differences between paired experimental and control classrooms within neighborhoods. (In other words, the experimental effect overrode variation among individuals but not among neighborhoods.) This phenomenon resulted because the treatment was effective in some neighborhoods but not in others. Indeed in certain neighborhoods, control groups also showed significant gains, although not as large as those achieved in the experimental classrooms.

Since each pair of classrooms was located in two schools in the same neighborhood, it occurred to Almeida that some feature of the neighborhood (such as school-community relations or ethnic tensions) might account for the differential effects. He therefore returned to each neighborhood to interview parents, teachers, and school personnel. In the course of this inquiry, Almeida discovered that the schools exhibiting greatest gains were located in neighborhoods having the most highly developed social networks, with the result that some experimental and control families were actually in communication with each other. Under these circumstances, not only the experimental classrooms but those of the control groups showed improvement, presumably as a function of horizontal diffusion. Indeed by the time Almeida returned to Mexico City for the follow-up interviews, he found that one or two parents from experimental neighborhoods were serving as leaders in a repetition of the parent participation program at the request and for the benefit of former control group families.

The most explicit and extended treatment of the impact of exosystems on the functioning of the school is provided by Ogbu's (1974) ethnographic study of interrelations between the school and other settings in the larger society. An anthropologist, Ogbu was asked to undertake such an investigation in connection with the introduction into the lower elementary grades of a bilingual (English and Spanish) instructional program. He defined his task broadly and attempted to understand the educational beliefs and behaviors of community residents, teachers, school administrators, and pupils, as well as to explain the high rates of school failure found in the community.

Burgherside is an unincorporated area within the city limits of Stockton, California. Its population is predominantly black and

Mexican-American. Its elementary school is part of the Stockton school system and is staffed by middle class professionals who do not live in the neighborhood. Its residents have low incomes; many were born in Mexico or the southeastern United States.

The children of Burgherside perform poorly in school by all available measures. Their scores on statewide tests are inferior, their report card grades are low, their dropout rate is high, and the proportion who continue their schooling beyond high school is below average. Ogbu notes three current explanations for school failure among poor and minority children: cultural deprivation (or difference), weaknesses in schools, and genetic inferiority. He rejects all three and proposes a fourth: that school failure is an adaptation to the limited opportunities for social and economic mobility available to members of "subordinate" minorities. In other words, educational inadequacy represents a response to discriminatory practices found in the larger society.

"Subordinate minorities," in Ogbu's terminology, include blacks, Mexican-Americans, and American Indians. They are distinguished from "immigrant minorities" who become Americans voluntarily. He points out that "subordinate" and "immigrant" minorities differ in their social status and in their typical school performance.

The author interviewed adult Burghersiders and their children, school personnel, and residents of contiguous neighborhoods, as well as middle class residents and community leaders of Stockton. He attended community meetings and social events regularly, observed in homes and schools, and reviewed school records. Information gathered in this way gave him a picture of the beliefs held by Burghersiders about education and of their actions in relation to those beliefs. It also revealed the conceptions of Burghersiders on the part of "taxpayers"—middle class Stockton residents, including school personnel.

These data led Ogbu to his conclusion that school failure is an adaptation to discrimination and attendant barriers to occupational and social achievement in adult life. He described this adaptation as having three components at the level of school and community: student's failure to perform at the highest possible levels, the patron-client relationship between school personnel and parents, and the definition of educational problems as clinical problems by school personnel.

The first of these components, students' failure to perform at maximum levels, is also Ogbu's explanation for at least part of the difference between the standardized test scores of subordinate minorities and those of other students. He argues that children in Burgherside, and others like them, simply do not attempt to maximize their scores on such tests. Neither do they try to get the highest possible grades in school. He cites interviews with students who professed satisfaction, almost universally, with "average" grades and who saw no reason to attend school regularly.

The school personnel Ogbu interviewed, and other middle class Stocktonians, often told him that attitudes such as these resulted from parents' low expectations for their children. But his own inquiries of those parents, and his measures of their children's aspirations, revealed quite high occupational and educational hopes. Furthermore, he witnessed and was told about many ways in which parents urged their children to perform well in school. Ogbu inferred from his interviews and observations of parent-child interactions that parents were communicating a double message. They did urge their children to succeed in school and to aspire to high status occupations, and the children tended to internalize those aspirations; but these parents also told their children that they were going to be victims of discrimination and that their most strenuous efforts would be frustrated. This second message was less consciously imparted than the first and was often embedded in comments or stories about discriminatory practices encountered by the parents themselves or by acquaintances. The effects of the second message were evident not only in school failure but in the widespread tendency of children to doubt that they would attain to the high goals they professed.

The label "taxpayers" that Ogbu applies to middle class Stocktonians was borrowed from their own characterizations of themselves, and it connotes much of what he means by the "patron-client relationship" between school personnel, representing middle class Stockton, and Burgherside parents. He employs the term ironically and demonstrates that a remarkably high percentage of Burgherside residents own their homes and therefore pay property taxes. The "taxpayers" are the people who make the important decisions about Burgherside's school and about the secondary schools attended by Burgherside youth. Schoolteachers, counselors, and administrators are uniformly identified with this group and see themselves as its representatives in Burgherside.

The patron-client relationship is revealed in many ways. Meetings are called in the school and other community centers to discuss "the problems of Burgherside." Residents do attend, but the talking is done by "taxpayers" who define the problem, explain what is to be done about it, and then inform the residents of what they will have to do to make the plan work. The Burghersiders do not express their real opinions, nor are they asked to do so. They leave feeling resentful, and little or no change occurs in their situation.

School personnel often cite poor communication as a factor in unsatisfactory school performance, but they implicitly define communication as a oneway process. Although the school sends information to Burgherside homes by way of notes, phone calls, and personal visits, parents are not given opportunities to state their concerns. Their resentment when an effective school principal was transferred to another school because he had done a good job in Burgherside was confined to the neighborhood because they were not consulted about school system actions. Their opinion that bilingual instruction would not help their children do better in school, especially the children from English-speaking families, was unheeded because of the attitude that such decisions are made elsewhere and the parents' role is to accept and support such decisions in whatever ways school personnel recommended.

Basic to the patron-client relationship between school personnel and parents, according to Ogbu, are "three functional myths." The first is that parent involvement in the school promotes academic success. He disposes of this myth empirically by comparing and finding no differences in the grades of students whose parents are and are not involved. He also describes cases in which changes in parent involvement had no effect on students' grades. Finally, he points out that no effects should be expected from parent involvement in the unequal relationship that prevails, except perhaps in the form of teachers feeling more sympathetic to children of parents who demonstrate proper deference.

The second myth is that Burgherside families have no fathers. Although the number of households headed by females is high, Ogbu finds that even fathers who are very concerned about their children's performance in school do not visit the school because among both black and Mexican-American males, involvement in the school is defined as woman's work. The third myth is that Burghersiders are caught in a welfare cycle extending over generations and that this cycle encourages high birth rates and illegitimacy and discourages good school performance. Ogbu shows this belief to be in conflict with the life histories of Burghersiders and their attitudes about welfare. Furthermore, a comparison of their children's school performance with pupils in Stockton whose families received welfare and with low income children not on welfare did not show any of the supposed differences.

The third component in school failure as an adaptive response to discrimination is, like the second, a reflection of the "taxpayers'" definition of the situation. Ogbu finds that school personnel, particularly guidance counselors, tend to define educational problems as clinical problems and to pursue clinical treatments at the expense of academic assistance. Guidance counselors see themselves as therapists and diagnosticians. They spend their time making referrals to clinics and in one-to-one therapeutic counseling rather than offering badly needed advice on routine academic matters such as course selection.

Among teachers, this tendency is expressed in the general attitude that the students are unable to learn because of their deprived backgrounds. The result is that standards are set low—teachers, like students, profess satisfaction with "average" performance—with the result that little differentiation is revealed among students or in the performance of the same students at different times. Ogbu traces report card comments on behavior and work habits over time to show that reported improvements in behavior are not reflected in higher grades.

The result of the definition of educational problems as clinical problems and of the patron-client relationship between "taxpayers" and Burghersiders is that various educational reforms are imposed on the community to counteract "cultural deprivation" without the understanding or agreement of the parents and without touching on the realities of discrimination and unemployment.

Ogbu's work bears directly on interconnections between school, family, and neighborhood and on the influence of economic conditions and community attitudes on school effectiveness. Moreover, he looks at the nature of the processes involved in these connections rather than merely at the statistical correlations between low income, minority status, and poor school performance. In this respect his work is fully consistent with the theoretical approach proposed here. His is truly an ecological study of school effectiveness at the level of both the meso- and the exosystems in which schools are embedded. His investigation is replete with findings supporting virtually all the hypotheses regarding direct and indirect connections between settings at both systems levels. In particular, Ogbu's results underscore the importance of bidirectional as against one-way communication, the existence of accurate information, mutual trust, positive feelings, and a sense of common goals in each group about the other, and above all a balance of power responsive to the needs of the developing person and to the efforts of those acting in his behalf.

Ogbu's research calls attention to several other properties of meso- and exosystems that have indirect but nevertheless profound consequences for human development. We are alerted to these properties by the dramatic contrast in social position within the community of the two principal groups on which Ogbu focuses his attention: the "subordinate minorities" on the one hand and the "taxpayers" on the other. It is clear that these groups belong to different social classes, but what makes the difference in their lives is not primarily the disparity in income, occupational status, education, or the places where they live. The critical factor is the settings that members of these groups do or do not enter and their position in these settings. The "taxpayers" were active in what I shall call *settings of power,* defined as settings in which the participants control the allocation of resources and make decisions affecting what happens in other settings in the community or in the society at large.

Power settings can be either formal (such as a board meeting) or informal (such as cocktail party or a golf game). They can occur at the local or the national level, either in the public (as in government) or the private sector (as in big business). The active participants in these settings, the persons who allocate the resources and make the decisions, are those whom C. Wright Mills called "the power elite" in the title and text of his classic work (1956).

As Ogbu so graphically shows, in Burgherside the settings and the seats of power are occupied by the "taxpayers." When "nontaxpayers" enter these settings, their position and the treatment they receive are rather different. Ogbu summarizes four of these differences:

(1) Public officials, especially those elected, speak of Taxpayers as the final arbiters of public policies and are careful to avoid offending them. (2) Taxpayers are the ones usually appointed to serve on various public boards and commissions, to act in the role of citizens. (3) When citizens are invited to express their views on public issues, the opinions of Taxpayers are considered more seriously than those of Nontaxpayers; this is true whether Taxpayers and Nontaxpayers speak as individuals or as representatives of their respective organizations—the organizations of Taxpayers having more influence on public policies than the organizations of Nontaxpayers. (4) Interests and opinions of Taxpayers receive

elaborate coverage in the local newspapers, radio, and television, whereas news coverage of Nontaxpayers occurs infrequently—only in relation to "their problems" with particular emphasis on their violation of the law and the various ways in which Taxpayers are helping them. (P. 51)

The general implications of this state of affairs for the ecology of human development are summarized in the form of a hypothesis pertaining to both meso- and exosytems.

HYPOTHESIS 44
The developmental potential of a setting is enhanced to the extent that there exist direct and indirect links to power settings through which participants in the original setting can influence allocation of resources and the making of decisions that are responsive to the needs of the developing person and the efforts of those who act in his behalf.

This hypothesis underscores the importance for human development of both first-order and second-order networks that connect the immediate settings containing the developing person with settings of power both in the local community and beyond. It follows that where these connections are remote, the effectiveness of the exosystem in promoting development is reduced accordingly. We incorporate this principle in a corollary hypothesis.

HYPOTHESIS 45
The developmental potential of a setting varies inversely with the number of intermediate links in the network chain connecting that setting to settings of power.

Ogbu's ethnographic study in effect documents the destructive consequences for human development that follow when the conditions stipulated in these exosystem hypotheses are violated. Such documentation carries with it powerful implications both for basic research and for public policy. For the science of development it emphasizes the need to explore the possible range and variety of exo- and mesosystem influences by going beyond the status quo to conceive and experimentally test intersetting arrangements not currently existing in the culture. This need is all the more desperate in that no society can long afford the systematic degrading and debilitation of large and richly talented segments of generations of its citizenry.

Both the Kohn and Ogbu researches deal not with unusual forms of interconnections between settings of special interest to the scientist but rather with patterns of class and caste that are to be found, to greater or lesser degree, in every American community. It is as if all the meso- and exo-structures under consideration had been constructed from a common blueprint—a phenomenon that signals the functioning of our final ecological structure: the macrosystem.

11.

The Macrosystem and Human Development

The macrosystem refers to the consistency observed within a given culture or subculture in the form and content of its constituent micro-, meso-, and exosystems, as well as any belief systems or ideology underlying such consistencies. Thus cultures and subcultures can be expected to be different from each other but relatively homogeneous internally in the following respects: the types of settings they contain, the kinds of settings that persons enter at successive stages of their lives, the content and organization of molar activities, roles, and relations found within each type of setting, and the extent and nature of connections existing between settings entered into or affecting the life of the developing person. In addition, these consistent patterns of organization and behavior find support in the values generally held by members of the given culture or subculture. In operational terms, the macrosystem is manifested in the continuities of form and content revealed by the analysis of a given culture or subculture with respect to the three prior levels of the ecological environment incorporated in our conceptual framework.

In terms of both its formal and substantive aspects, this ecological conception of the macrosystem contrasts sharply with the implicit theoretical model underlying the prevailing research approach to the study of macrosystem influences on human development: the identification of class, ethnic, and cultural differences in socialization practices and outcomes. The typical strategy in such inquiries is to select a sample of children or parents from contrasting backgrounds and then to document the observed differences in methods of child rearing and/or their effects. With few exceptions, the former are assessed by questionnaires and the latter by test results or experimental procedures. Given the concept of a macrosystem

258

proposed here, such data are useful but hardly sufficient. Some systematic information is required regarding the structure and substance, at least at one ecological level, of the environments in which the reported behavior took place. To state only that the research subjects are Swiss and American parents or children from low income rather than middle class families is to provide but a marker, a sign on the door of an environmental context that leaves its nature unspecified. Under such circumstances any inference about process, which is the main concern of scientific inquiry, becomes little more than speculation. Such investigations are not without value, but they tend to pose many more questions than they answer. For this reason, in my judgment, they no longer, if they ever did, represent a strategy of choice for research on human development. Since I have contributed more than my share to such endeavors, it is only fitting to cite a study of which I was a coauthor to illustrate the shortcomings, along with a few strengths, of this widely employed strategy in the study of development-in-context.

Research carried out by a combined team of Israeli and American investigators (Kav-Venaki et al., 1976) had as its basic aim to study the developmental effects of an ecological transition from a Communist authoritarian to a Western democratic society. The sample consisted of forty one Russian-born adolescents from Jewish families that had emigrated to Israel. These youngsters participated in an experiment in which they reported their readiness to engage in morally disapproved-of behavior such as denying responsibility for property damage or cheating on a test. Social pressure was created by telling the children, after giving them an initial base condition, that their responses would be shown either to their parents and teacher or to their classmates. In a counterbalanced design, instructions and questionnaires were presented once in Russian and once in Hebrew.

The results were compared with those from previous studies involving Russian school children in Moscow (Bronfenbrenner, 1967) and native-born Israeli pupils in Tel Aviv schools (Shouval et al., 1975). They revealed that, regardless of the language of administration, levels of conformity for the émigré children fell between those for Soviet and Israeli youngsters but were closer to those of the latter. Within the sample, the longer a child had lived in one or the other society, the more her response to social pressure resembled the modal reaction of children in that society. Children from families who, while still in the U.S.S.R., had spoken Yiddish in the home showed marked differences in response from those from families who had not, with the former resembling the Israeli and the latter the typical Soviet reaction. Contrary to our hypotheses, when subjected to pressure from adults the émigré children gave more conventional moral responses when instructions were given in Hebrew rather than Russian. This result was interpreted as reflecting the tendency to respond more moralistically to the language of authority, which for the émigré children had shifted from Russian to Hebrew. In general the findings indicated that young people who had been exposed through early adolescence to the authoritarian regime characteristic of Soviet classrooms nevertheless exhibited, after only two years of residence in a Western country social reactions much closer to those of native-born Israeli children. It was impossible to determine from the available data, however, to what extent the response of the émigré children had its origins in a Jewish family upbringing while they were still living in the U.S.S.R., as against subsequent exposure to the Israeli environment.

On the basis of these findings, we drew the following conclusion.

Our results testify to the responsiveness and flexibility of the child in adapting to radical changes in the context and process of socialization. Both field and experimental studies have documented the intensity and univocal character of Soviet methods of collective upbringing and their power to evoke conformity among Russian schoolchildren (Bronfenbrenner, 1967, 1970a, 1970b). Yet, our findings indicated that young people who had been exposed to these methods through adolescence showed, after only 2 yr's residence in another country, a markedly contrasting commitment to values of autonomy and independence. The residues of the earlier Soviet orientation were still present, but the characteristic Israeli response to social pressure was clearly stronger. To be sure, some of our data also indicated that the now prevailing dissident direction had its origins in family upbringing while the children were still living in the U.S.S.R., but that fact only reinforced our evidence for the capacity of the child to adapt to conflicting socialization settings both within and across contrasting cultures. (Kav-Venaki et al., 1976, p. 85)

Intriguing as these findings and interpretations may be, they permit virtually no conclusions about the contrasting environments involved in the ecological transition or the children's process of adaptation after arrival in the new setting. Investigation of these phenomena would have required a different methodology relying more on observation and interviewing than on questionnaires and employing experimental situations that are real rather than hypothetical (such as working on a group project in the presence or absence of the teacher).

It is ironic to present research on hypothetical situations in a discussion advocating rigorous, ecologically validated studies of class and cultural differences. But in this domain, by contrast with other areas of inquiry, we have not been able to find close approximations to the requirements of an ecological model with respect to either systematic assessments of the environment or measures of developmental outcome.

There is one area of research on macrosystem influences on development which, though as yet sparsely settled, employs a much more differentiated conception of the external environment than is found in the populous terrain of class and cultural studies. These are investigations of social change and its effects on psychological growth.

Although the first research to be considered uses conventional outcome measures in the form of psychological tests and laboratory-type procedures, this limitation is more than compensated for by the breadth and brilliance of its treatment of the process of development. In addition, the study contributes a whole new dimension to our ecological conception of the macrosystem. The work may be excused for having some traditional features, since it was conceived and conducted almost half a century ago, by one of the most creative researchers in the discipline of psychology. But because the work clashed in its theoretical orientation and substantive findings with the prevailing scientific and political ideology of the time, its publication was delayed for over forty years.

The circumstances under which the investigation was undertaken are described by the principal investigator, the late Russian psychologist A. R. Luria, in a preface to the American edition (1976) of the first Soviet publication of his research monograph two years earlier.

The history of this book is somewhat unusual. All of its observational material was collected in 1931–32, during the Soviet Union's most radical restructuring: the elimination of illiteracy, the transition to a collectivist economy, and the readjustment of life to new socialist principles. This period offered a unique opportunity to observe how decisively all these reforms effected not only a broadening of outlook but also radical changes in the structure of cognitive processes.

The Marxist-Leninist thesis that all fundamental human cognitive activities take shape in a matrix of social history and form the products of sociohistorical development was amplified by L. S. Vygotsky to serve as the basis of a great deal of Soviet psychological research. None of the investigations, however, was sufficiently complete or comprehensive to verify these assumptions directly. The experimental program described in this book was conceived in response to this situation, and at Vygotsky's suggestion.

We did our research in the remoter regions of Uzbekistan and Kirghizia, in the *kishlaks* (villages) and *dzhailaus* (mountain pasturelands) of the country. Our efforts could have met with equal success, however, in the remoter areas of European Russia, among the peoples of the North, or in the nomad camps of the Siberian Northeast. Despite the high levels of creativity in science, art, and architecture attained in the ancient culture of Uzbekistan, the masses had lived for centuries in economic stagnation and illiteracy, their development hindered among other things by the religion of Islam. Only the radical restructuring of the economy, the rapid elimination of illiteracy, and the removal of the Moslem influence could achieve, over and above an expansion in world view, a genuine revolution in cognitive activity.

Our data indicate the decisive changes that can occur in going from graphic and functional—concrete and practical—methods of thinking to much more theoretical and abstract modes brought about by fundamental changes in social conditions, in this instance by the socialist transformation of an entire culture. Thus the experimental observations shed light on one aspect of human cognitive activity that has received little scientific study but that corroborates the dialectics of social development. (Pp. v–vi)

Luria does not allude in his volume, however, to the circumstances requiring the four-decade delay in publication. These are described in a foreword by the editor of the American edition, a distinguished psychologist in his own right, Michael Cole.

After two expeditions during which the data in this book were gathered, Luria made some preliminary public descriptions of his results, but the intellectual climate in Moscow at the time was not at all friendly to his conclusions. Although Luria clearly emphasized the beneficial consequences of collectivization, critics pointed out that his data could be read as an insult to the people with whom he had been working (Razmyslov, 1934). The status of national minorities in the USSR has long been a sensitive issue (not unlike the issue of ethnic minorities in the United States). It was all well and good to show that uneducated, traditional peasants quickly learned the modes of thought characteristic of industrialized, socialist peoples, but it was definitely not acceptable to say anything that could be interpreted as negative about these people at a time when their participation in national life was still so tenuous. (P. xiv)

The theoretical orientation of Luria's work anticipates in several respects the convergence of ecological and developmental perspectives that I have set forth here. Luria's position is summarized in Cole's introduction.

Part of the initial controversy over Luria's cross-cultural work may have arisen from the developmental orientation he brought to this topic. His general purpose was to show the sociohistorical roots of all basic cognitive processes; the structure of thought depends upon the structure of the dominant types of activity in different cultures. From this set of assumptions, it follows that practical thinking will predominate in societies that are characterized by practical manipulations of objects, and more "abstract" forms of "theoretical" activity in technological societies will induce more abstract, theoretical thinking. The parallel between individual and social development produces a strong proclivity to interpret all behavioral differences in developmental terms. (Pp. xiv–xv)

Ever the experimentalist, even when on horseback in remote villages in Soviet Asia, Luria organized his fieldwork within the framework of an experimental design. He took advantage of the existence of naturally occurring experimental and control groups resulting from the fact that the revolution had not yet fully penetrated the Islamic areas of the country.

Naturally enough, these regions of the Soviet Union were undergoing especially profound socioeconomic and cultural changes. The period we observed included the beginnings of collectivization and other radical socioeconomic changes as well as the emancipation of women. Because the period studied was one of transition, we were able to make our study to some extent comparative. Thus we could observe both underdeveloped illiterate groups (living in villages) and groups already involved in modern life, experiencing the first influences of the social realignment.

None of the various population groups observed had in effect received any higher education. Even so, they differed markedly in their practical activities, modes of communication, and cultural outlooks. (Pp. 14–15)

Luria's final sample consisted of five groups ranging from collective farm workers at one extreme to traditional peasants still relatively untouched by the revolution at the other. The outcome measures consisted principally of tests of cognitive functioning in such classical areas as perception, deduction and inference, and problem solving, but they also included assessments of capacity for imagination and level of self-evaluation.

The results of the study are best conveyed in Luria's own words.

The facts show convincingly that the structure of cognitive activity does not remain static during different stages of historical development and that the most important forms of cognitive processes—perception, generalization, deduction, reasoning, imagination, and analysis of one's own inner life—vary as the conditions of social life change and the rudiments of knowledge are mastered.

Our investigations, which were conducted under unique and non-replicable conditions involving a transition to collectivized forms of labor and cultural revolution, showed that, as the basic forms of activity change, as literacy is mastered, and a new stage of social and historical practice is reached, major shifts occur in human mental activity. These are not limited simply to an expanding of man's horizons, but involve the creation of new motives for action and radically affect the structure of cognitive processes . . .

Closely associated with this assimilation of new spheres of social experience, there are dramatic shifts in the nature of cognitive activity and the structure of mental processes. The basic forms of cognitive activity begin to go beyond fixation and reproduction of individual practical activity and cease to be purely concrete and situational. Human cognitive activity becomes a part of the more extensive system of general human experience as it has become established in the process of social history, coded in language.

Perception begins to go beyond graphic, object-oriented experience and incorporates much more complex processes which combine what is perceived into a system of abstract, linguistic categories. Even the perception of colors and shapes changes, becoming a process in which direct impressions are related to complex abstract categories . . .

Together with new forms of abstract, categorical relationships to reality, we also see the appearance of new forms of mental dynamics. Whereas before the dynamics of thought occurred only within the framework of immediate, practical experience and reasoning processes were largely limited to processes of reproducing established practical situations, as a result of the cultural revolution we see the possibility of drawing inferences not only on the basis of one's own practical experience, but on the basis of discursive, verbal, and logical processes as well . . .

All these transformations result in changes in the basic structure of cognitive processes and result in an enormous expansion of experience and in the construction of a vastly broader world in which human beings begin to live . . .

Finally, there are changes in self-awareness of the personality, which advances to the higher level of social awareness and assumes new capabilities for objective, categorical analysis of one's motivation, actions, intrinsic properties, and idiosyncracies. Thus a fact hitherto underrated by psychology becomes apparent: sociohistorical shifts not only introduce new content into the mental world of human beings; they also create new forms of activity and new structures of cognitive functioning. They advance human consciousness to new levels.

We see now the inaccuracy of the centuries-old notions in accordance with which the basic structures of perception, representation, reasoning, deduction, imagination, and self-awareness are fixed forms of spiritual life and remain unchanged under differing social conditions. The basic categories of human mental life can be understood as products of social history—they are subject to change when the basic forms of social practice are altered and thus are social in nature.

Psychology comes primarily to mean the science of the sociohistorical shaping of mental activity and of the structures of mental processes which depend utterly on the basic forms of social practice and the major stages in the historical development of society. The basic theses of Marxism regarding the historical nature of human mental life are thus revealed in their concrete forms. This becomes possible as a result of the radical, revolutionary shifts permitting us to observe, over a brief period, fundamental changes which under ordinary conditions would require centuries. (Pp. 161–164)

To place Luria's thesis in the context of our conceptual framework, the macrosystem also undergoes a process of development and in doing so lends movement to all its composite systems down to the level of the person. Thus the members of a changing society necessarily experience developmental change at every psychic level —intellectual, emotional, and social.

Luria's conception represents a kind of analogue, in the science of human development, of Einstein's relativity principle in physics. Just as Einstein shattered the Newtonian view of motion as a departure from a fixed reference point, so Luria requires us to conceive of individual development as occurring within a dynamic environmental system. To corrupt a metaphor from Einstein's explanation of his Special Relativity Theory: development takes place in a moving train, and that train is what we may call the "moving macrosystem."

If there are two trajectories, one embedded within the other, what is the relation between them? Is the individual simply caught in the current of history, or does he exhibit a momentum of his own? How much lag is there? Does the past leave its mark on the present? And for how long?

These are not questions that can be answered by Luria's research, for two reasons. First, the answers to these questions require observations from at least two points in time, and Luria took only one reading. The absence of follow-up was not the result of lack of opportunity or practical difficulties; it was based on a deliberate decision. In the concluding paragraphs of his book, written as a contemporary addendum to the original manuscript, he states:

Scholars who took upon themselves the task of examining our work as it was being prepared frequently expressed the wish that we carry out the same research again in order to make a comparative analysis of the further changes that have occurred over the past forty years in these locations. While this suggestion is quite reasonable, we do not feel compelled to follow it.

Our data show what major changes in the structure of cognitive processes began to take place during the period of our original research, shifts that had already taken place in the first years of the cultural revolution for the inhabitants of the remoter parts of our country. Since then, the author has repeatedly been to Uzbekistan and has witnessed the enormous changes in social and cultural life that have occurred during these years. To repeat the research in the same localities forty years later, during which time the peoples of central Asia have, in effect, made a leap of centuries, would be superfluous. An investigator who desired to replicate our work would obtain data that differ little from those he might obtain by studying the structure of cognitive processes among inhabitants in any other part of the Soviet Union. (P. 164)

Luria's moving statement reflects the second and more compelling reason why his work does not speak to our expressed concerns. His primary interest, at least in this particular study, was in a group phenomenon occurring in reaction to a historical event at a given point in time, not in the development of the individual over an extended period. To obtain answers to our questions, we shall have to look to research over a longer time span that conceives of the person, in his relation to the forces of social and economic change, as actor as well as object.

Fortunately, such an investigation does exist. Again the "independent variable" was a great social and economic upheaval, but on this occasion a distinctly American one, the Great Depression of 1929.

In 1962, Glen Elder, a research sociologist working at the Institute of Human Development at the University of California at Berkeley, recognized a unique scientific opportunity. Forty years earlier, the institute had initiated, under the leadership of its directors, Harold Jones and Herbert Stolz, a major longitudinal study involving a large sample of eleven-year-old children in Oakland, California. The project was later extended through periodic follow-up investigations from preadolescence to midlife. In 1964, Elder's endpoint in the present analysis, the subjects were in their forties, most of them with families of their own.

The scientific opportunity that Elder recognized, and subsequently realized (1974), was a built-in research design created by a natural, albeit societal, catastrophe. The families of these children had all lived through the period of the Great Depression, but, as happens in a mighty storm, some had been directly hit by the disaster and others spared. Moreover, as far as individual households were concerned, these strokes of fortune fell in a virtually random pattern as this factory closed and that one remained in operation, one stock collapsed and the other survived on the Big Board.

Elder undertook to compare the life course development of those who had, and had not, been exposed to the full force of the Depression. And since as a sociologist he had escaped the training of the training of the laboratory researcher and its individualistic bias, Elder chose as the object of his investigative concern not just the child but the total family.

He also transcended the confines of his own discipline, which in its preoccupation with the complexities of societal and institutional structure occasionally loses sight of the individual human being in process of development. Elder managed to lose sight of neither the growing forest nor its separate growing trees. As Clausen has written in his foreword to Elder's monograph,

> In periods of crisis, the element of chance seems to play a major role in influencing life outcomes. At such times, we can hardly specify an expectable life course beyond the immediate impact of the crisis. The task of delineating "net effects," of tracing out the various patterns of impact, response, and ultimate influence, seems almost insuperable. Only by combining historical, sociological, and psychological perspectives with detailed, longitudinal data on individual experiences, orientations, and behaviors can such an analysis be accomplished. This is precisely what Glen Elder has done in the present volume. (P. xv)

Like the phenomenon he was studying, Elder's theoretical orientation and basic method were also distinctively American, although they departed from the mainstream preoccupation with behavioristic models and large-scale computer analysis of "objective" data. The acknowledged prototype for Elder's own sociohistorical investigation was the pioneering, monumental work in this same domain by W. I. Thomas, a sociologist of the Chicago school. In Elder's words,

> The problem before us is not simply whether economic change produced family and generational change, or the nature of that change; it includes questions concerning the process by which such change occurred. What are the conceptual linkages between economic change and the adult

careers of men and women who were children in the 30s? It may be clear from what has been said that basic features of this approach are indebted to the early work of William I. Thomas, and especially to his classic study (with F. Znaniecki) *The Polish Peasant in Europe and America* (1918–20). Thomas trained his analytic eye on linkages between social structure and personality and made a convincing case for studying such linkages at points of discontinuity or incongruence between person and environment, as seen in his theory of crisis situations, of adaptations to new situations. From the vantage point of the present study, we also appreciate Thomas's emphasis on developmental concepts of life experience, on the use of life records and histories. (P. 7)

One additional aspect of Thomas's orientation permeates Elder's approach—his insistence that the social world can be understood only by analyzing how it is experienced by those who live in it. In this connection Elder quotes Thomas's view that social science "must reach the actual human experiences and attitudes which constitute the full, live and active social reality beneath the formal organization of social phenomena ... A social institution can be fully understood only if we do not limit ourselves to abstract study of its formal organization, but analyze the way in which it appears in the personal experience of various members of the group and follow the influence which it has upon their lives" (p. 338).

Elder had at his disposal extensive data on each family, including lengthy interviews with parents, teachers, and the child, as well as staff observations and ratings of social and emotional behavior, self-report questionnaires, personality inventories, and a psychiatric assessment. All these materials were examined and evidence extracted for information bearing on patterns of stability and change in the family's relations with and position in the community, the functioning of the family as a system, and the behavior and experience of the offspring, first as children living with their parents, attending school, and spending time with their peers, and then as adults with families of their own, participating in the worlds of work, social life, and civic affairs.

Elder's experimental design involved comparing patterns of stability and change in two groups differentiated on the basis of whether loss of income as a result of the Depression exceeded or fell short of 35 percent. This dividing line between the so-called deprived and nondeprived families was based on a previous finding that "families lost or were forced to dispose of assets with some frequency only when income losses exceeded 40 percent" (p. 45). Such reduction in income as was sustained by the nondeprived group was appreciably relieved by the marked decline in cost of living during the Depression era. Both groups were further dichotomized by the family's social class (middle or working) before the Depression, resulting in a two-by-two design.

The results of Elder's systematic analysis and the conclusions he reached about the influence of the Depression experience on family functioning and the development of the child, are most conveniently summarized in the two general categories of short- and long-term effects.

As might be expected, the immediate effects of severe economic loss were reflected in symptoms of emotional distress reported by parents, particularly those from working class families. The strain experienced by the parents was also visible to the children. Both boys and girls from deprived families were more likely to check the item "I wish my mother (father) were happier." A second, more salient difference was observed in the emerging dominance of the mother in families hit hardest by the Depression. This phenomenon was often accompanied by a lowered status of the father in the eyes of the children, whereas the mother's perceived importance increased.

In Elder's view, these shifts in intrafamily dynamics are a consequence of the perceived "role failure on the part of the husband" and the resultant shift of economic responsibility to the mother and other family members (p. 28). The circumstances were, of course, not of the father's own making. In Elder's words,

It is not surprising that male victims of the Depression were frequently blamed for set-backs over which they had little or no control, given a value system that extols individual responsibility and self-sufficiency. More important, however, is the generalized acceptance of this utilitarian self-evaluation among husbands and fathers who were deprived of the means to adequately support their families. Instead of attributing cause to deprivational conditions in society, to a force beyond the individual actor and his understanding, the record, such as it is, shows that the unemployed or hard-pressed workers were inclined to direct hostile feelings and frustrations toward the self, punishing themselves for the consequences of an economic system. (P. 104)

Again, the Thomases' harsh maxim prevails: "If men define situations as real, they are real in their consequences" (1928, p. 572).

In addition to showing greater preference for the mother over the father, children from deprived families expressed a stronger identification with the peer group. Indeed Elder emphasizes that, in the sphere of social relations, *"the attractiveness of age-mates stands as the most significant effect of economic hardship."* The difference was evidenced primarily in the desire to have many friends rather than a few, since "close friends were just as common among children in both deprivational groups according to their report." Although the trend was apparent for both sexes, "orientation toward friends is strongest among boys from deprived homes" (p. 97).

But perhaps the most salient characteristic of "children of the Depression" was their greater participation in domestic roles and outside jobs, with girls specializing in the former, and boys in the latter. By the time they were teen-agers, about 90 percent of the daughters from deprived families were doing domestic chores compared with 56 percent for the nondeprived. Among the boys, 65 percent of those from deprived as against 42 percent among the nondeprived families were doing some kind of paid work. For sons in families in which the father was unemployed, the boys' part time employment rate rose to 72 percent. The job percentages for the girls showed a similar trend but with lower participation rates.

In Elder's view the greater involvement of children from deprived families in domestic and economic roles has developmental implications.

One consequence of a decremental change in family status and resources is to heighten children's awareness of parental investments which made possible the goods and services they had formerly taken for granted. These include the effort and skills which provide income for the family unit, as well as the labor involved in homemaking and child care. Economic scarcity brought out the reciprocal aspects of consumption which entail obligations to others. Especially in middle-class families, deprivation generally changed one-sided dependency regimes, in which parents indulged their offspring's desires, to an arrangement where children were expected to demonstrate more self-reliance in caring for themselves and family needs. (P. 66)

Consistent with this interpretation, children from deprived families, and particularly children whose fathers were unemployed, were rated as more adult-oriented, defined by such descriptions as "seeks adult company, hangs around adults making frequent bids for attention, identifies with adults and is very cordial to them" (p. 81).

Early experience in jobs appeared to have special impact on the motivation of boys from deprived families.

Awareness of family hardship is implied in the outlook of children who felt deprived of spending money and desired greater control over their life situation. For boys, in particular, gainful employment is a logical out-let for these motivational orientations in situations of economic hardship, and the data generally show this connection between family deprivation and economic activity ... Likewise, boys with a deprived background and those with a job were most likely to be described as ambitious in social aspirations during the high school years; trained clinicians rated them highest on the desire to control their environment by suggestion, persuasion, or command. (P. 70)

In light of the findings on the behavior of children from homes hardest hit by the Depression, Elder concludes as follows: "Accord-ing to our analysis, the roles children performed in the economy of deprived families ... oriented them toward adult ways. Economic hardship and jobs increased their desire to associate with adults, to 'grow up' and become an adult" (p. 82).

The above results and conclusions have social as well as scientific significance: it is probably no coincidence that in today's society, as work opportunities for school-aged children and youth have reached new low levels, school achievement has been decreasing at the same time that vandalism and violence in the schools, along with juvenile delinquency in the streets, have been on the rise (Bronfenbrenner, 1976, 1978b).

The fact that longitudinal data were available over a period of more than three decades placed Elder in the unique position of being able to assess the impact of childhood experience, within and outside the family, on behavior in later life. The long-term conse-quences of the Great Depression turned out to be rather different for men and women and were further qualified by the social status of the family at the time when it was hit by economic hardship.

The findings for males run counter to the expectations based on conventional wisdom: sons whose families were hardest hit by the Depression profited by the experience.

In both the middle and working classes, boys from deprived families arrived at a firmer vocational commitment in late adolescence and were more likely to be judged mature in vocational interests than the offspring of nondeprived parents. In adulthood, they entered a stable career line at an earlier age, developed a more orderly career, and were more likely to have followed the occupation which they preferred in adolescence. Vocational maturity in crystallized interests established a positive link between family deprivation and occupational attainment, and at least partly offset the educational handicap of a deprived background. Desire to excel in adolescence proved to be the most important source of occu-pational achievement for boys from deprived families in both social classes, and was highly related to mental ability. (Pp. 200–201)

These trends were more marked for men who had grown up in families of middle class status before the Depression. For them economic misfortune had an especially salutary effect. Not only were they more successful occupationally but they were also "judged more favorably on psychological competence and health than any other group in the sample" (p. 248). Outside of work they were family-centered, preferring family activities to leisure-time pursuits or community involvement and viewing children as a major source of gratification.

The picture in adulthood for working class boys whose families were victims of the Depression is not as bright. Though no less motivated than their deprived counterparts from middle class homes, they were often unable to obtain higher education. Perhaps for this reason, they showed somewhat more evidence of psychological disturbance and had the highest percentage of heavy drinkers (43 percent) of any group.

Women whose families suffered economic loss reflected their childhood experiences in adult life by emphasizing the maternal role that had been so prominently enacted by their mothers.

> Our data suggest that receptivity to traditional roles is concentrated among women who grew up in deprived households that depended heavily upon the involvement of female members. From adolescence and the late 30s to middle age, a domestic life style is more characteristic of these women than of women from nondeprived homes. They were more involved in household chores, expressed greater interest in domestic activities, and, in the middle class, were more likely to marry at an early age. A smaller proportion entered college, when compared to the non-deprived in each social class ... but a larger percentage married into the upper middle class ... the daughters of deprived families were most likely to stop working at marriage or when they gave birth to their first child; and (if from the middle class) to enjoy the common tasks of homemaking. The meaning of family preference centered first on the value of children and secondarily on the interpersonal benefits of marriage. (P. 239)

Adults whose families had escaped economic ruin present a paradox. Compared with their counterparts whose parents had experienced the full force of the Depression, they were less successful both educationally and vocationally; a greater proportion were judged to have psychiatric problems, including heavy drinking (43 percent versus 24 percent). Elder comments on what he calls "their surprising resemblance in symptoms and adaptive deficiencies to working class offspring" despite the fact that "they were a privileged group in the Depression." He then poses a question and provides his own answer. "Why are these adults not the healthiest, most competent members of the Oakland cohort? It seems that a childhood which shelters the young from the hardships of life consequently fails to develop or test adaptive capacities which are called upon in life crisis. To engage and manage real-life (though not excessive) problems in childhood and adolescence is to participate in a sort of apprenticeship for adult life" (pp. 249–250).

In the words of the banished Duke, "Sweet are the uses of adversity" (*As You Like It*). But in keeping with Thomas's thesis we should ask, "From whose point of view?" The children in the Oakland sample were born in 1920–1921, so that by the time the full impact of the Depression had struck, they were old enough to understand and contribute to meeting the grave difficulties their families were experiencing. What would have happened had they been younger and still dependent practically and psychologically on their parents, who were then facing overwhelming problems of their own?

As a true disciple of Thomas, Elder has dealt with this question, and has done so empirically. Together with a colleague (Elder and Rockwell, 1978) he has analyzed data from another longitudinal study, conducted over the same period in the more affluent community of Berkeley, California but with a sample of males born in 1928–1929. These children spent their earliest years during the depth of the Depression. The authors state their principal research aim as follows:

Our objective in this study is to carry out a more restricted longitudinal analysis of the Depression experience in the life patterns and health of 83 males who were born in Berkeley, California (1928–29), an investigation which will nevertheless permit selected *intercohort* comparisons of this historical change in social experience and psychological functioning. By comparing outcomes from the Berkeley cohort with those obtained from similar analyses of the Oakland men (birthdates, 1920–21), we shall be able to test a *life stage* hypothesis; *that Depression entailed more adverse and enduring developmental outcomes in the lives of men who encountered this event as young children than as adolescents*. The Oakland youth left home for school, work, and marriage at the end of the 1930s; the Berkeley adolescents at the end of World War II. The latter were more vulnerable than their Oakland counterparts to family strains and disruptions in the Depression and they were exposed to a longer phase of economic hardship and its persistence up to departure from home. The psychological significance of this difference in timing is suggested by a young child's dependence on significant others that were unpredictable, sullen and often hostile (Pp. 251–252).

Drawing on the findings of the earlier study, Elder and Rockwell formulated two additional hypotheses qualifying the predicted effects of economic loss as a function of the family's social class status and the stage of life after which the sequelae of deprivation would be most pronounced. In the former sphere, because of their more vulnerable position boys from working class backgrounds were expected to be more adversely affected in their development by economic hardship than their middle class counterparts. With respect to timing, the developmental impact of the Depression was anticipated to be greater during the formative years, extending from childhood through the completion of formal education, than in the subsequent period from leaving school to midlife, a phase structured primarily by participation in the world of work.

As the dividing line between families experiencing more versus less deprivation during the Depression, Elder and Rockwell used the same criteria as had been employed in the previous investigation—35 percent loss in income. When this information was not available, as occurred for a substantial number of families, resort to unemployment and public assistance records produced supplementary samples of deprived and nondeprived families in the same proportion as that obtained for groups classified by income loss. Hence the samples were pooled.

The comparison of these two groups on measures of developmental progress yielded results that are consistent with Elder and Rockwell's principal hypothesis regarding the greater vulnerability of younger children to the stresses generated by economic hardship. The most striking evidence is the sharp contrast of these results with those of the earlier Oakland study. Whereas the family's financial distress had a positive effect on the life course development of middle class boys who experienced the Depression as adolescents, the very opposite occurred for boys who had lived through their early childhood during this period.

For example, sons from both middle and working class whose families had suffered income loss ranked lower on high school grades and aspirations than the nondeprived, and they were less likely to complete college (34 percent versus 64 percent). Thus for the young child, the Depression controlled a major gateway to success in the adult years. Even among boys who did enter college, an analysis of clinical ratings based on all the case materials available revealed the developmental costs of a deprived background. Especially in the middle class, "it is youth who grew up in deprived households who are characterized more by a lack of self-esteem and personal meaning in life, by a tendency to withdraw from adversity, avoid commitments, and employ self-defeating tactics; and by a sense of victimization and vulnerability to the judgment of others" (p. 271). Youth not in college did not show these differences in significant degree. Failure to enter college increased the negative consequences of family deprivation for the occupational careers of the Berkeley men. In this educational category, men from deprived families entered the labor force at an earlier age than the nondeprived, spent a larger proportion of their worklife in manual jobs, and exhibited a more unstable work pattern, as indicated by multiple employers and shifts between different lines of work. In the final follow-up at about age forty, a higher proportion of the deprived than nondeprived men reported (table 9) "some health problem or impairment" (64 percent versus 41 percent), "chronic fatigue or energy decline" (46 percent versus 20 percent), and "heavy or problem drinking" (44 percent versus 26 percent). This trend was particularly marked among deprived men who became high achievers (G. Elder, personal communication).

The more pronounced psychological effects of Depression hardship among the sons of middle class parents contradict Elder and Rockwell's second hypothesis on class variations. In retrospect, the authors suggest that deprived families at this socioeconomic level are a sub-group "most subject to discontinuity between scarcity conditions and the aspirations of wartime America" (p. 274). A complementary explanation appears in the analysis of the same phenomenon in the Oakland study. Noting that parental distress did not vary as strongly by economic deprivation in the working class, Elder comments, "Intense economic need among all working class families, the lower visibility of status decline, and social comparisons with the middle class may account for this result" (1974, pp. 62–63). To risk an awkward metaphor, at the bottom of the social ladder, one runs into a floor effect. The same mechanism may account for the finding in the Berkeley study that deprivation-linked differences in psychological characteristics observed among men with college experience did not appear at lower educational levels.

The findings for the third hypothesis, regarding the stage of development at which effects of economic loss would be strongest, were more surprising, more complex, and perhaps for these reasons more instructive. Although the boys entered first grade at a time when the Depression was still a source of family stress, there was

no evidence of this strain in the school performance, emotional state, and social relationships during grade school . . . neither tests scores nor psychological development varied by economic deprivation within each [social class] stratum . . . The developmental costs of family hardship *only* [italics in original] emerge during the boys' adolescent years in World War II, a period of rising prosperity . . . In retrospect, this finding corresponds with the proposition that developmental limitations are likely to surface when children encounter demanding situations that call upon their adaptive resources, as in the transition from the protective environment of an elementary school to the achievement pressures of adolescence. (Elder and Rockwell pp. 270, 274.)

As to developmental effects in the period after adolescence, we have already taken note of the disturbed work patterns and psychological problems exhibited by men from deprived homes, especially those who did not go on to college.

If they entered college, however, as over 60 percent of those in the Berkeley sample did, "men with deprived histories were more likely than the nondeprived to embark on a course that produced substantial worklife achievement, and they did so without the advantage of more years of education or any detectable asset through adolescence" (p. 281).

This pattern contains a paradox that does not escape the investigators.

Men observed in adolescence as unambitious, submissive, and indecisive do not come to mind as workers who quickly established themselves career-wise and stayed with their line of work over an extended period of time. On the contrary, one would expect to see evidence of floundering and vacillation, a disorderly work pattern of frequent job shifts, periods of idleness, and status fluctuations. How, then, do we connect two phases within a life course that appear to have so little in common? How did a good many of the sons of deprived parents manage to achieve occupational success from a background of educational limitations and maladaptive behavior in adolescence? (P. 281)

The authors' first step in answering these questions was to demonstrate that the successful career patterns characterizing the college entrants among the deprived group were accompanied by evidence of positive change in other domains of psychological functioning as well. For this purpose, they dichotomized the men in this group into those who had attained high occupational status relative to their education (high achievers) and those who had not (low achievers), were selected to be "identical on measured intelligence, high school grades and aspirations, and even level of formal education" (p. 279).

To assess directions and differences in development, Elder and Rockwell compared personality ratings (Q sorts) administered at adolescence, thirty, and forty years of age. The overall trend was "toward greater confidence and health," but the high achievers from deprived backgrounds showed the greatest positive change on such variables as "high aspiration level for self," "genuinely values intellectual and cognitive matters," "low self-esteem," "indecision," and "withdrawal from adversity." In other respects, however, such as impulse control, consideration for others, and resilience in the face of criticism, they were at a disadvantage relative to their non-deprived counterparts.

Considering the deprived group as a whole, Elder and Rockwell sum up their developmental status at age thirty as follows:

By the age of 30, they are much less likely to resemble their adolescent portrait of low self-esteem, indecision, and withdrawal from adversity than are the nondeprived who also moved upward in worklife. At this stage, children of deprived families remain more vulnerable to the judgments of others, when compared to the nondeprived, but they are no longer distinguished by feelings of inadequacy and meaninglessness, by self defeating behavior and reluctance to commit self to a course of action. In these respects at least, we see evidence of relatively greater inner strength, effectiveness, and purpose than was observed in adolescence ... The developmental course of successful men from deprived families is characterized by a mixture of strengths and weaknesses, of developmental gains since adolescence and persistent deficits; and some pathogenic traits are more prominent at middle age than in early adult years, e.g., lacks personal meaning, self defeating. Nevertheless, they have accomplished far more than one would have expected from their early background and lives. (P. 287.)

The picture even improved somewhat during the next ten-year period, but the psychological costs of economic stress were still in evidence.

The years between age 30 and 40 brought greater satisfaction and well-being to men from deprived families in general, owing partly to their accomplishments, and they were more likely than the nondeprived to claim that life had improved since the troubled years of adolescence. But the benefits of worklife achievement did not completely eliminate the health risk of family hardship by middle age—the emotional pain and detachment, confusion and insecurity. The healthiest Berkeley men are among those who combined the support of a nondeprived family with the rewards of adult achievement; at the other extreme we find men whose worklife bears some resemblance to family misfortune in the Great Depression; that is, the sons of deprived parents who were relatively unsuccessful in their worklife. (P. 299.)

What circumstances accounted for the positive changes after adolescence in persons suffering economic deprivation during the formative years? Elder is currently pursuing the answer to this question through further analyses of data from both the Oakland and Berkeley Studies, and some indication is already given of the emerging trends. "Our analysis underscores the value of military service, a rewarding worklife, and possibly the emotional support and gratifications of marriage and family life. In the successful group, men from deprived families were no less the beneficiaries of a stable, satisfying marriage than the offspring of more privileged family backgrounds" (p. 290).

At the same time Elder and Rockwell warn that the available data might not be equal to the task. In speaking of the "developmental gains" exhibited after adolescence by sons of families experiencing severe economic loss, they state: "This change may have been prompted by adult independence and work, or by marriage, parenthood, and 'psychosocial moratorium' of military service. The tendency to drop out of college for work or military service could be interpreted as symptomatic of floundering; and also as valuable maturing experience that enabled some men to achieve a sense of direction and purpose in life. The precise source of change will be difficult to establish, given the materials at hand" (pp. 281–282).

The original question that prompted Elder and Rockwell to replicate the Oakland analyses with the Berkeley was: would the impact of the Great Depression be more damaging to the development of younger children than of adolescents? The authors summarize their answer, and their evidence, in the following excerpts from their conclusion:

Family deprivation produced greater disadvantage for the life course and health of the Berkeley men up to middle age, when compared to the Oakland cohort; an effect most pronounced during the early years of adolescence and formal education . . . Unlike the Oakland cohort, family deprivation is associated with lower adolescent aspirations and school performance among the Berkeley men, with impaired self-esteem, vulnerability to the evaluations of others, indecision and passivity—with actions that generally enhanced their difficulties.

Through economic and developmental constraints, Depression hardship restricted life prospects by curtailing higher education among the Berkeley men, regardless of class origin, and it did so to a greater extent than in the Oakland cohort . . .

Comparison of the Berkeley and Oakland cohorts thus identifies both similarities and differences in the life outcomes of Depression hardship. Similarities include the concentration of developmental handicaps during the early years of adolescence and formal education, the effectiveness of worklife experience in countering educational limitations, and the perception that life has become more satisfying since adolescence. Differences center on the relatively young age of the Berkeley men when hard times occurred and on their more prolonged exposure to such conditions; from adolescent development to higher education and health at mid-life, family deprivation entailed more adverse outcomes for these men than for their older counterparts in the Oakland cohort. It is among the Berkeley offspring of deprived families that we see the greatest discontinuity between childhood and adult experience, and their significant contributions to psychological well-being in middle age. The legacy of family deprivation remains a problem at mid-life even among the most successful men. (Pp. 298–299.)

In a subsequent report, Elder (1979) presents findings on the daughters of families from the Berkeley sample and compares the results with those obtained in the Oakland study. Once again the analysis yields a paradoxical result. Whereas, as we have seen, boys whose early childhood occurred during the Depression were adversely affected in their development, the effect on girls, though much less marked, was in the opposite direction. From early on, "the Berkeley girls fared well in deprived families and appear more goal oriented, self adequate, and assertive in adolescence than the daughters of nondeprived parents . . . Despite the strains and privations of deprived households, most Berkeley women emerged from this family history as competent, resourceful adolescents."

Elder traces the origin of this differential outcome for the two sexes to the contrasting childhood experiences of daughters and sons in families subjected to severe economic strain. He cites evidence that "boys in deprived families *lost more* in affection for father and *gained less* in warmth toward mother when compared to girls" resulting in "a weaker tie between father and son and a much stronger tie between mother and daughter."

An additional factor interacted with these asymmetrical relationships to influence markedly the impact of economic stress on sons and daughters in the Berkeley sample. Using data from interviews conducted before 1929, Elder derived an index of the quality of the marital relationship. The index turned out to be an important predictor of the effect of the Depression on subsequent development, but only for boys. Elder summarizes his findings, and the presumed underlying process, in the following passage:

Economic loss among the Berkeley families weakened the tie between father and son while increasing the solidarity of mother and daughter. However, both outcomes were contingent on the pre-Depression marital bond. When inadequate as family earner, father's relation to son and daughter was most heavily dependent on the strength of the marital bond and the wife's support. Under conditions of marital harmony, economic loss actually enhanced the relationship of father with son and daughter. In the absence of this pre-condition, deprived girls acquired closer ties with mother, when compared to the nondeprived, whereas deprived boys generally lost emotional support in relation to mother and especially father.

It is important to bear in mind that all the processes and effects set in motion by the Great Depression were age-specific. Of particular interest is the fact that children who were victims of the Depression as adolescents profited from the experience for years to come; their development was enhanced through the life span as a result of their exposure to economic deprivation. There is a lesson to be learned here, and Elder emphasizes it.

The labor-intensive economy of deprived households in the 30s often brought older children into the world of adults . . . These children had productive roles to perform. But in a more general sense they were needed, and, in being needed, they had the chance and responsibility to make a real contribution to the welfare of others. Being needed gives rise to a sense of belonging and place, of being committed to something larger than the self. However onerous the task may be, there is gratification and even personal growth to be gained in being challenged by a real undertaking if it is not excessive or exploitative. Thus we are not referring here to the desperate situation of many Depression children who lived a life not unlike that of the children of Mayhew's London poor—the offspring of the costermongers who worked the streets of mid-nineteenth-century London selling fruit, vegetables, and fish (Mayhew, 1968). For most of the Oakland children in deprived families, expecially in the middle class, productive status in the household economy did not require an educational sacrifice or even a noteworthy limitation on social contacts with age-mates. Our point is that economic losses changed the relation of children to the family and adult world by involving them in needed work which contributed to the welfare of others. Much of this work entailed "people services," in contrast to gainful employment. Similar change is noted in the vast literature on families and communities in natural disasters; the young often play vital roles in the labor-intensive emergency social system.

Since the Depression and especially World War II various developments have conspired to isolate the young from challenging situations in which they could make valuable contributions to family and community welfare. Prosperity, population concentration, industrial growth with its capital-intensive formula, and educational upgrading have led to an extension of the dependency years and increasing segregation of the young from the routine experiences of adults. In this consumption-oriented society, urban middle-class families have little use for the productive hands of offspring, and the same applies to community institutions . . .

This society of abundance can and even must support "a large quota of nonproductive members," as it is presently organized, but should it tolerate the costs, especially among the young; the costs of not feeling needed, of being denied the challenge and rewards which come from meaningful contributions to a common endeavor? (1974, pp. 291–293)

I have given considerable attention to Elder's two complementary studies of "children of the Great Depression" because his work adds substantially to the understanding of the power and possibilities of an ecological model in research on human development. His research has produced evidence bearing directly on many of our already existing hypotheses, both within and beyond the microsystem. Furthermore, the imaginative scope of the data he reports, and the thoughtful interpretation he provides, reveal still other relationships and possibilities that can guide future investigations. For this reason, it is profitable to review Elder's work from the perspective of our theoretical framework in relation to each of four levels of the ecological schema.

The two settings that figure most prominently in Elder's analyses are the home and the workplace. Two others receive significant but secondary attention—the school and the peer group. A good deal is revealed about processes in each setting, and some of it corroborates our earlier propositions about the operations of microsystems as they affect psychological growth. Elder's findings regarding the developmental impact of children's participation in household chores and part time work underscore the importance of the child's engaging in molar activities (hypothesis 1) and in a diversity of roles (hypotheses 9 through 10) to enhancing both motivation and competence. Elder's analysis of the effects of early participation in economic roles by children of deprived families underscores the importance for the child's development not only of engaging in new roles and new types of molar activity but also of early involvement with adults outside the family.

HYPOTHESIS 46
The development of the child is enhanced through her increased involvement, from childhood on, in responsible, task-oriented activities outside the home that bring her into contact with adults other than her parents.

As in the case of earlier propositions dealing with optimal conditions for development, this hypothesis is subject to the qualification that the balance of power governing such responsibilities gradually shifts in the direction of the child so as not to inhibit her evolving capacity of initiative and creative contribution to the task at hand.

More dramatically, Elder's demonstration of the decisive part played by the structure of relations between mother, father and child in mediating the impact of economic stress testifies to the crucial significance of three-person systems in shaping the course of human development (hypothesis 8). If this system had been stable prior to the Depression, not only were family and child able to survive its onslaught but the experience of stress actually had salutary effects. If the intrafamilial linkages were weak, the economic blow too heavy, and the children still young, the concomitant denigration of the father and impairment of his capacity to function in a parental role had, as we have seen, profound effects on the future development of the children in the family, particularly boys. The three-person system can be likened to a three-legged stool. When all three legs are balanced, there is maximal stability, and it takes a heavy blow to create an upset; but if one leg is damaged or broken, the arrangement becomes precarious and the stool may collapse.

But in some instances the extreme stability of the system can impede individual development. The decisive factors here become the adequacy of the support previously provided and the person's age and sex. By providing examples of both kinds of developmental disruption, Elder's findings constitute a vivid illustration of the ecology of human development in operation: "the progressive, mutual accommodation between an active, growing human being and the changing properties of the immediate settings in which the developing person lives" (definition 1). The striking differences in Elder's data for the effects of the Great Depression on teenagers versus preadolescents and boys versus girls reflect the influences of sex and biological maturation in this process of mutual accommodation.

The family is not the only context in which changing biological forces affect the structure of the emerging psychological field. Another is highlighted by Elder's conclusion, based on findings from the Oakland study, that "the attractiveness of age-mates stands as the most significant effect of economic hardship" (1974, p. 97). This factor may have played a more powerful role in shaping the lives of "Children of the Great Depression" than is reflected in Elder's analysis. The trend found strongest expression in a preference for many friends as opposed to a few and was particularly marked for boys. For the Oakland sample this additional impetus to peer involvement brought on by the Depression did not occur until adolescence. But if, as seems likely, the Depression had similar effects in nearby Berkeley, then the boys in the sample would have been exposed to peers from an earlier age and without the probably counterbalancing influence of a respected male model in the person of a successful father—a model that had been available prior to adolescence for sons in the Oakland study. Significant in this regard is the general finding from studies of children from father-absent homes that they tend to be more susceptible to peer group pressure and to exhibit a pattern of behavior characterized by low motivation for achievement and low self-esteem, leading eventually, under the influence of the peer group, to greater impulsiveness and aggression. The differences remain after control for social class (for a summary and references see Bronfenbrenner, 1961; Hetherington, Cox, and Cox, 1977). These traits are not dissimilar to those that differentiated deprived and nondeprived youth in the Berkeley sample.

Among boys who continued their education beyond high school, we find considerable evidence in adolescence of a developmental pattern shaped by the Depression experience . . . Whatever their class origin before hard times, it is youth who grew up in deprived households who are characterized more by a lack of self-esteem and personal meaning in life; by a tendency to withdraw from adversity, avoid commitments, and employ self-defeating tactics; and by a sense of victimization and vulnerability to the judgments of others . . . The least responsible adolescents were characterized as opportunistic, inclined to stretch limits, and less able to control impulses; a behavior pattern which persists into the adult years among the economically deprived. (Pp. 270–271, 286.)

Finally, we have seen in research reviewed previously that, at least in the United States, the peer group tends to undermine adult socialization efforts and to encourage egocentrism, aggression, and antisocial behavior.

All this evidence suggests that increased exposure to peer group influences, as an indirect effect of growing up in the Depression from early childhood, may have contributed to the impaired educational, vocational, and psychological development exhibited by the adult males in the Berkeley sample. To check on this speculation, one would need to relate the extent of peer involvement among deprived and nondeprived children to their subsequent life course.

The definition of developmental ecology is not limited by any single setting; it accords equal importance to relations between settings and to the large contexts in which the settings are embedded. In its implications for these broader domains Elder's work provides support for earlier ecological hypotheses and a basis for generating new ones. In an age when the organization of data bits into supposedly meaningful clusters is all too often left to a preprogrammed computer, Elder painstakingly examines and orders his facts to reveal how the structure of one setting, and the child's experience in it, generate expectations and action patterns that are in part carried over to and in part radically transformed by experience in another setting that invites or even compels other kinds of activities, roles, and patterns of interpersonal relation.

Surely the most spectacular outcome of Elder's work is his demonstration that events in one setting exert their influence on a person's competence and relations with others in quite another setting decades later. Just as Luria recognized the crucial significance of the time dimension for the macrosystem, so Elder demonstrates the temporal elongation of exo- and mesosystem connections. Experiences in one setting carry over into other settings, often over extended periods of time.

In Elder's research, the settings that were most important in this regard were the family and the peer group, and they are likely to be so in every life course because of their special properties. The nature of these properties is specified in the next hypothesis.

HYPOTHESIS 47

The developmental potential of a setting is a function of the extent to which the roles, activities, and relations occurring in that setting serve, over a period of time, to set in motion and sustain patterns of motivation and activity in the developing person that then acquire a momentum of their own. As a result, when the person enters a new setting, the pattern is carried over and, in the absence of counterforces, becomes magnified in scope and intensity. Microsystems that exhibit these properties and effects are referred to as *primary settings,* and the persisting patterns of motivation and activity that they induce in the individual are called *developmental trajectories.*

The most pervasive and potent primary settings in human societies are, of course, the family and the workplace, although the power of the latter to generate what I have called developmental trajectories is only now being demonstrated systematically, principally through the work of Kohn. A close third is the peer group, although its unstable and short-lived character limits its impact to the extent that other more enduring settings remain a prominent part of the person's life.

It is the operation of these primary settings behind the scenes of Elder's drama of life careers that mediates the external impact of economic forces to produce the impressive continuities revealed by his masterful analysis. In particular, he has illuminated the mechanisms maintaining and even strengthening earlier developmental trajectories through a succession of ecological transitions, first from home to school, and then from school to either continued education or direct entry into the world of work.

Before considering possibilities for departures from an established course of development, we must understand the conditions that set in motion and tend to perpetuate an established trajectory. Elder has demonstrated that the effects of processes occurring in one setting may not be observable until the person enters some other setting later in life.

Under what circumstances are such sleeper effects likely to occur? In the Berkeley sample, the effects of early economic deprivation did not become manifest until adolescence. In Elder's view the critical factor accounting for this phenomenon was "the transition from the protective environment of the elementary school to the achievement pressures of adolescence" (p. 274). It seems likely, however, that the young person's relations with other settings— such as family and peer group—were undergoing similar changes at the same time. The issue may be stated more generally in the form of a hypothesis.

HYPOTHESIS 48
Developmental effects are not likely to be manifested until the person moves from his present primary setting into another, potential primary setting, that is, from a setting that has instigated and currently maintains the person's present level and direction of functioning to another setting requiring the person to take initiative to find new sources of stimulation and support. Such transition between two primary settings is called a *primary transition*. Sleeper effects of earlier primary settings are most likely to be observed after primary transitions have taken place, since these are usually separated in time by months or years.

Here is another example of a now familiar principle: to demonstrate that an ecological trajectory has been developed it is necessary to show that it carries over and persists in a new setting.

This hypothesis has implications for method, substance, and public policy. On the methodological side, the proposition represents a restatement of our criterion for developmental validity (definition 9). It implies that the enduring developmental effects of a setting cannot be effectively assessed within that same setting. This statement, in turn, carries a substantive implication: as long as a person remains in the same primary setting, one cannot know with any assurance whether that setting is having a beneficial or baneful influence on the person's psychological growth; the behavior observed may be merely adaptive and not reflect any genuine developmental change. From the perspective of public policy, in the absence of appropriate research studies we are in danger of tolerating long-standing environmental situations that may in fact be harmful to psychological growth, and the undesirable effects remain undetected as long as people remain in these situations. For example, pupils in large schools on the outskirts of town "don't look all that bad." In similar fashion, situations conducive to growth may be overlooked because of sleeper effects; for instance, children in a home visiting program "don't seem to be anything special."

Perhaps the most revolutionary implication of Elder's findings is that patterns not only of continuity and decline but also of recoupment and even renaissance can be seen occurring over time in mesosystems. Data to suggest recovery with time also emerged from the studies of the long-range effects of institutionalization, reviewed earlier. But Elder's results, because they deal with the mesosystem, enable us to identify some of the ecological conditions that can bring about substantial change in the development of the person well beyond the childhood years.

The clearest evidence in Elder's work for this phenomenon comes from the analysis of the precursors and sequelae of college attendance in the lives of his subjects. In the Berkeley study, the Depression experience was a powerful predictor of whether the student graduated from college. Among those college entrants whose family had suffered severe economic loss when they were young, 43 percent "made it" compared with over 80 percent for the nondeprived.

But it was not graduation from college that predicted the later occupational career of Depression victims; what mattered was college entrance, that is, some exposure to higher education after high school. "Having entered college, men with deprived histories were more likely than the nondeprived to embark on a course that produced substantial worklife achievement" (Elder and Rockwell, p. 281). As it turned out, neither exposure to college nor subsequent achievement on the job spared them the frustrations of an unstable path to success or the psychological legacy, in adulthood, of "emotional pain and detachment, confusion and insecurity (p. 299)."

The fact remains, however, that for young men who were victims of the Depression in their childhood years, the ability to enter college, whether or not they managed to graduate, made a big difference in the rest of their lives. Entrance into college was not a guarantee for the future, but, as Elder and Rockwell's analyses show, it provided opportunities completely closed to those who did not matriculate.

College entrance is of course strongly correlated with socioeconomic status. But if this factor were controlled for, what personal or social circumstance would determine whether or not a young person matriculated? The Berkeley study does not provide any data regarding this question. All we know is that early economic deprivation did not make the difference, since the rates, of college entry for deprived and nondeprived men were exactly the same. Undoubtedly some of the decisive factors had their roots in characteristics of the individual. But others were surely external, a function of conditions and events unrelated to the qualifications of the person. Under these circumstances, whether a given opportunity is available becomes a critical feature of the ecological field, setting the course of the person's future development. Hence our next hypothesis.

HYPOTHESIS 49
The direction and degree of psychological growth are governed by the extent to which opportunities to enter settings conducive to development in various domains are open or closed to the developing person.

It is instructive to examine this same phenomenon in greater detail as it emerges from Elder and Rockwell's data. The critical factor appears to lie in the properties of the settings that the young person enters after leaving a Depression-stricken home. In particular, improvement is associated with college entry, marriage, military service, and the work experience. Except for an allusion to Kohn's analysis of the work setting as a context for development, Elder and Rockwell do not discuss the qualities of the new settings or the circumstances of the transition that make the difference—again, in all likelihood, for lack of the necessary evidence. On the basis of the other research here reviewed, however, and the theoretical propositions it has generated, the following hypothesis seems justified.

HYPOTHESIS 50
The developmental effect of a transition from one primary setting to another is a function of the match between the developmental trajectory generated in the old setting and the balance between challenge and support presented both by the new setting and its interconnections with the old. The nature of this balance is defined by previous hypotheses specifying the conditions of micro-, meso-, and exosystems conducive to psychological growth, with due regard to the person's stage of development, physical health, and degree of integration with as opposed to alienation from the existing social order.

A final feature of the extension of the mesosytem through time merits attention. The developing person, upon entering into new settings, participates in new roles, activities, and patterns of interrelationship. If the basic assumption underlying our entire theoretical approach is valid, such expanding participation sets the necessary conditions for human development to take place. I have defined human development as "the process through which the growing person acquires a more extended, differentiated, and valid conception of the ecological environment, and becomes motivated and able to engage in activities that reveal the properties of, sustain, or restructure that environment at levels of similar or greater complexity in form and content." The definition by no means implies that mere entry into a new setting is an indication that development has occurred. But a person cannot maintain a role, engage in role-appropriate activities, and sustain a pattern of ongoing interpersonal relations in a setting without being motivated or without acquiring "a more extended, differentiated, and valid conception of the ecological environment" (definition 7). Hence once such activities have occurred, some development has already taken place. The question of how much is an empirical one. As I stated at the outset, activity is at once the source, the process, and the outcome of development. The extent to which it occurs in an ever-expanding ecological environment thus becomes the measure of developmental progress. From this point of view, the involvement of the person in human activity in a succession of new settings represents a developmental trajectory in the making.

In a somewhat similar theoretical context, Freud affirmed as the guiding principle of human development, "Where id was, there shall ego be" (1933, p. 112). Freud's formulation was entirely intrapsychic, literally "out of this world." By contrast, an ecological orientation is interactive and very much in this world: in it, development involves making the world one's own and becoming a person in the process. If there is an ecological analogue to Freud's psychoanalytic injunction, it is much less elegant but has the pragmatic advantage of being workable. As befits an ecological model, it is couched in systems terms: "Where exo- is, there shall meso- be." In other words, the developing individual begins to move into and to master those segments of the external environment that control his life.

But as I have steadfastly maintained, at the heart of an ecological process there is always a two-stage sequence. What about the missing link? If Freud left out the superego from his formula, that is no reason for an ecologist to omit the "super" level of the ecological system from a final formulation. This level is that of the macrosystem.

Although hardly referred to explicitly, the macrosystem has in fact been the persistent theme of this final chapter. I have spoken only of micro-, meso-, and exosystems, but the addition of the phrase "as they do, or do not exist today in American society" makes the entire discussion completely on target. For the status quo is what the macrosystem is about—but not all about. In keeping with Leontiev's law (see chapter 2), the macrosystem encompasses the blueprint of the ecological environment not only as it is but also as it might become if the present social order were altered. Moreover, transforming experiments necessarily involve the macrosystem since they represent efforts to achieve "the systematic alteration and restructuring of existing ecological systems in ways that challenge the forms of social organization, belief systems, and lifestyles prevailing in a particular culture or subculture" (definition 11).

Elder's historical analysis is not without significance for the alteration and restructuring of existing ecological systems, since it does challenge a dominant belief system that pervades American society in general and, in particular, both research and social policy in the field of human development. To a substantial degree, the prevailing ideology underlying these professional activities is imbued with a "deficit model" of human function and growth. Such a model assumes that what we view as inadequacy or disturbance in human behavior and development—even, or perhaps especially, when it is not the product of organic damage—reflects some deficiency within the person or, from a more enlightened but fundamentally unaltered perspective, within that person's immediate environment. One begins with the individual, looking for signs of apathy, hyperactivity, learning disabilities, defense mechanisms, and the like. If this attempt is not successful, one knows just where to look next. If the source of the deficiency is not to be found within the child, it must lie with the parents: they aren't providing the child with enough cognitive stimulation, they haven't worked through their relationship to one another, or their personalities are still fixated at a preoedipal level. (The possibilities are endless; the chief target of our social service programs across the land is multiproblem families.) And if the source of difficulty remains elusive, the ethnic or social group to which the family belongs can always be blamed. There must be something wrong with somebody, and somebody usually turns out to be the person or group having the problem in the first place. The presumed task of professionals, be they researchers or practitioners, is to find the deficiency and do their best to correct it but without hoping for too much: after all, that's the way those people are; they do not really want to change.

The above indictment is strong and sweeping. But there can be no doubt that these practices are widespread in our culture. One need only count the number of professionals and auxiliary personnel, both in the private and public sectors, who are employed specifically for the purpose of diagnosing the deficiencies presumed to lie within the person or his family and of carrying out corrective procedures, again within the same constricted domain, albeit with little hope of effecting a significant, lasting improvement.

As for research, one has only to survey the scientific publications in the field, particularly those dealing with development-in-context, to recognize the deficit model often underlying the choice of problems, variables, methods, and research design. Indeed even in the present volume, in which a conscious effort was made to right the balance by seeking out as examples investigations that would at least permit the documentation of the active, constructive, cooperative potential of the species, the deficit model is implicit in several of the studies occupying a prominent place in the argument, notably the "Eichmann experiment", Zimbardo's Pirandellian prison, and Moore's comparative study of the long-range effects of day care. All of them emphasize the darker and presumably more dominant and enduring aspect of human nature and its development.

It is precisely this presumed locus, primacy, and persistence of human frailty that are challenged by Elder's finding that many children of the Great Depression and their families, though suffering patent psychic damage, struggled, overcame, and in some instances actually profited from the trauma of sudden poverty and learned helplessness in an unresponsive environment. To be sure, some sequelae remained, as they will always remain as long as scientists, practitioners, and policymakers in the field of human development continue to be resigned to the status quo.

The alternative is the rejection of a deficit model in favor of research, policy, and practice committed to transforming experiments. The purpose of such transforming endeavors is twofold. First, they serve the goals of science by implementing Dearborn's dictum: "If you want to understand something, try to change it." But why challenge, alter, and restructure the existing social order if not to make a more human ecology—to create new micro-, meso-, and exosystems that better meet the needs of human beings and then, if they work, to write these systems into a revised societal blueprint? Here is the missing link that constitutes the second half of the final formulation. It is fitting that it be couched as a three-person system that exhibits both stability and momentum. The ecologist's injunction is to love, honor, and perhaps even to obey, Dearborn's dictum, Leontiev's law, and a new version of Thomas's thesis: "Experiments created as real, are real in their consequences."

NOTES

BIBLIOGRAPHY

INDEX

Notes

2. Basic Concepts

1. This does not mean that these settings have no place in ecological research. On the contrary, I argue that laboratory experiments are powerful and often essential tools for illuminating the distinctive properties of a given ecological environment provided that the laboratory results are complemented by relevant data from other settings.

2. In a recent survey of all studies in child development ($N = 902$) published between 1972 and 1974 in three prominent research journals (*Child Development, Developmental Psychology,* and *Journal of Genetic Psychology*), Larson (1975) found that 76 percent of all the investigations had employed the experimental laboratory paradigm; the next highest category was research using pencil-and-paper techniques (17 percent); observational studies were in last place (8 percent).

3. The Nature and Function of Molar Activities

1. One society that does so is the People's Republic of China. For a description see Kessen (1975).

4. Interpersonal Structures as Contexts of Human Development

1. For a comprehensive review of this literature, see Lamb (1976a).

2. Parsons and Bales (1955) have provided a detailed analysis of the properties of such a four-person family structure as a context for socialization, but the work is essentially theoretical and neither supported by nor directed toward empirical research.

3. In general, social network theorists (such as Bott, 1957; Mitchell, 1969) have used the term *density* to describe the extent to which interconnections exist between members, but there are no exact counterparts to what we differentiate as open versus closed systems. The distinction is important for the ecology of human development in view of what emerges as the geometrically increased power of closed social networks as contexts for socialization, especially at the level of the meso- and exosystem (see chapters 9 and 10).

5. Roles as Contexts for Human Development

1. The investigators describe the prisoners' self-depreciation as a tendency "to adopt and accept the guards' negative attitude toward them" (p. 86). The description calls to mind Bettleheim's statement, in an account of his own experiences as a prisoner in Nazi concentration camps that, in the final stages of imprisonment, inmates "seem to have a tendency to identify with the Gestapo," even to the point of sewing and mending their uniforms "so that they would resemble those of the guards" (1943, p. 448). Indeed Bettelheim's analysis of the psychological changes that he observed over time in his fellow prisoners bears a striking resemblance to the developmental course described by Zimbardo and colleagues for inmates of their simulated prison. Like their Stanford counterparts, concentration camp inmates initially "tried to react . . . by mustering forces which might prove helpful in supporting their badly shaken self-esteem" (p. 428) but ultimately became apathetic and submissive, showing a "child-like dependence on the guards" that turned them into "more or less willing tools of the Gestapo" (pp. 444, 447). The similarity between the reactions of college students who had agreed only to play the role of prisoner in a "make-believe" jail and the behavior of their counterparts in a tragic prison situation constitutes additional evidence for the ecological validity of the Stanford experiment.

2. Again there is a striking parallel between the behavior of prisoners and guards in a simulated prison and in the stark reality of a concentration camp. Thus Bettelheim reports: "It seemed to give pleasure to the guards to hold the power of granting or withholding the permission to visit the latrines . . . This pleasure . . . found its counterpart in the pleasure the prisoners derived from visiting the latrines because there they could causally rest for a moment, secure from the whips of the overseers and guards. They were not always so secure, because sometimes enterprising young guards enjoyed interfering with the prisoners even at these moments" (1943, p. 445).

6. The Laboratory as an Ecological Context

1. Physiological data consistent with these findings had been reported in an earlier experiment utilizing modification of the strange situation in both laboratory and home, with heart rate as a dependent variable (Sroufe, Waters, and Matas, 1974).

2. The confounding was deliberate on the investigator's part and based on the typical instructions given by investigators when conducting research on mother-infant interaction in the two settings. Belsky's intent was to demonstrate the danger of generalizing from studies conducted in one setting to behavior in another.

3. The role of the unfamiliar in enhancing distress reactions in young

children is nicely illustrated in a study of the one-year-old's reaction to separation from the mother in the home setting (Littenberg, Tulkin, and Kagan, 1971). The children exhibited significantly greater anxiety when the mother departed not through the usual exit door but through one that was rarely used.

4. Four additional studies comparing the reaction of day care versus home-reared children to maternal separation are qualified by special circumstances. Two of these investigations revealed no significant setting effects. The first, by Kagan, Kearsley, and Zelazo (1978), was a well-controlled study, but departed from the standard Strange Situation procedure by eliminating the introduction of a stranger, thus making the experience less stressful. Cochran (1977) did introduce the stranger, but conducted the experiment in the home. Of the remaining two experiments, Doyle and Somers (no date), like Moskowitz, Schwarz, and Corsini (1977), found greater distress among the home-reared infants, but the comparison groups were poorly matched. Only Ricciuti (1974) obtained results consistent with Blehar's hypothesis of greater vulnerability for infants in day care, but he called his own findings into question because of a possible methodological artifact resulting from inconsistency in the procedures followed with the two groups of children. In any case, the differences obtained in Ricciuti's sample of only nine infants were not statistically significant.

5. An extensive literature in clinical psychology and psychiatry does, of course, focus to a large extent on the subjective experiences of the person in the socioemotional realm. This focus is typically restricted, however, to the sphere of interpersonal relations and seldom encompasses larger social structures, particularly those beyond the level of the microsystem. Moreover, the subject's experience is usually reported and interpreted solely as it pertains to a preconceived theoretical structure (for instance, psychoanalysis) and is seldom investigated in the framework of a rigorous research design, let alone a systematically planned environment.

6. Weisz justifiably criticizes researches conducted outside the laboratory for their tendency to collect data that lead only to what he calls "empirical statements" as opposed to theoretical propositions. Examples are reports about observed differences in developmental studies of children (or the behavior of their caretakers) associated with a variety of ecological contrasts, as between home and day care, one and two parents, kibbutz and family upbringing, age desegregated and homogeneous classrooms, working and middle class families, American and Russian child rearing and so on. Findings of this sort certainly do not tell us very much beyond the descriptive level. Their banality derives not from the fact that the data were gathered in naturalistic settings, as Weisz seems to imply, but from the atheoretical nature of the research

questions. Surely laboratory studies have contributed their full share to the store of statistically significant but substantively sterile findings that all too often survive editorial review to populate our scientific journals.

9. Day Care and Preschool as Contexts of Human Development

1. One possible explanation for this variation has been proposed by R. Darlington, co-director of the Lazar project. In a personal communication, he suggested that the effect of the experimental programs in reducing retention may have been attenuated because, in some school systems, the two outcome measures are confounded; pupils who might have otherwise been left back are placed in special programs instead.

2. These findings can be misinterpreted in terms of their implications for public policy. The present *Federal Interagency Day Care Requirements* (Department of Health, Education, and Welfare, 1968) mandate caretaker-child ratios of one to five in centers with children three to four years of age and of one to seven for four- to six-year-olds. Since the low ratio group in the Atlanta experiment was still within the federal guidelines, the experiment provides no information on effects in centers that violate these guidelines. Yet it may be that precisely beyond this limit a reduced child-caretaker ratio begins to have substantial deleterious impact. Liberalization or revocation of the guidelines would doubtless have the practical effect of decreasing the number of caretakers per child, since in the *National Day Care Study* it was found to have "the most substantial impact on cost per child of any single factor studied" (p. 48).

9. The Mesosystem and Human Development

1. Hayes and Grether's findings have recently been replicated by Heyns (no date) with a sample of almost fifteen hundred sixth-graders in the Atlanta public schools. As in the New York City study, the gap between low and middle income children, and between black and white pupils, widened disproportionately during the summer months.

10. The Exosystem and Human Development

1. Indexed by agreement or disagreement with such assertions as: "the most important thing to teach children is absolute obedience to their parents", "there are two kinds of people in the world: the weak and the strong", and "in this complicated world, the only way to know what to do is to rely on leaders and experts" (Kohn, 1969, p. 79).

2. The raw data were actually broken down into steps, such as "some grade school," "grade school graduate," "some high school," and so on. From the perspective of an ecological model, it would have been desirable (and quite feasible) to examine the relative impact of each setting and transition on value change.

Bibliography

Acton, J. E. 1948. *Essays on freedom and power.* Boston: Beacon Press.

Ahrens, R. 1954. Beitrag zur Entwicklung der Physionomie und Mimiker-kennens. *Zeitschrift für Experimentelle und Angewandre Psychologie* 2:412–454; 599–633.

Ainsworth, M. D. 1962. The effects of maternal deprivation: a review of findings and controversy in the context of research strategy. In *Deprivation of maternal care: a re-assessment of its effects.* W. H. O. Public Health Papers, no. 14. Geneva: World Health Organization.

Ainsworth, M. D. S., and Wittig, B. A. 1969. Attachment and exploratory behavior of one-year-olds in a strange situation. In *Determinants of infant behavior,* vol. 4, ed. B. M. Foss. London: Methuen.

Ainsworth, M. D. S., and Bell, S. M. 1970. Attachment and exploratory behavior of one-year-olds in a strange situation. *Child Development* 41:49–67.

Ainsworth, M. D. S., Bell, S. M. V., and Stayton, D. 1971. Individual differences in strange situation behavior of one-year-olds. In *The origin of human social relations,* ed. H. R. Schaffer. London: Academic Press.

Aldrich, C. K., and Mendkoff, E. 1963. Relocation of the aged and disabled: a mortality study. *Journal of the American Geriatrics Society* 11:185–194.

Almeida, E. 1976. An experimental intervention for the development of competence in Mexican sixth grade children. Doctoral dissertation, Cornell University.

Ambrose, J. A. 1961. The development of the smiling response in early infancy. In *Determinants of infants behavior,* vol. 1, ed. B. M. Foss. New York: John Wiley.

Aronson, E., and Carlsmith, J. M. 1968. Experimentation in social psychology. In *The handbook of social psychology,* vol. 2, ed. G. Lindzey and E. Aronson. Reading, Mass.: Addison-Wesley.

Asch, S. E. 1956. Studies of independence and conformity: a minority of one against the unanimous majority. *Psychological Monographs* 70:no. 9 (whole no. 416)

Avgar, A., Bronfenbrenner, U., and Henderson, C. R. 1977. Socialization practices of parents, teachers, and peers in Israel: Kibbutz, Moshav, and city. *Child Development* 48:1219–1227.

Baldwin, A. L. 1947. Changes in parent behavior during pregnancy. *Child Development* 18:29–39.

Bales, R. F. 1955. Adaptive and integrative changes as sources of strain in social systems. In *Small groups: studies in social interaction,* ed. P. Hare, E. F. Borgatta, and R. F. Bales. New York: Alfred A. Knopf.

Banuazizi, A., and Movahedi, S. 1975. Interpersonal dynamics in a simulated prison: a methodological analysis. *American Psychologist* 30:152–160.

Barker, R. G., and Gump, P. V. 1964. *Big school, small school.* Stanford, Calif.: Stanford University Press.

Barker, R. G., and Schoggen, P. 1973. *Qualities of community life.* San Francisco: Jossey-Bass.

Barker, R. G., and Wright, H. F. 1954. *Midwest and its children; the psychological ecology of an American town.* Evanston, Ill.: Row, Peterson.

Barnett, C. R., Leiderman, P. H., Grobstein, R., and Klaus, M. 1970. Neonatal separation: the maternal side of interaction deprivation. *Pediatrics* 45:197–205.

Bayley, N. 1932. A study of the crying of infants during mental and physical tests. *Journal of Genetic Psychology* 40:306329.

Bee, H. L., Van Egeren, L. F., Streissguth, A. P., Nyman, B. A., and Leckie, M. S. 1969. Social class differences in maternal teaching strategies and speech patterns. *Developmental Psychology* 1:726–734.

Bell, S. M. 1970. The development of the concept of the object and its relationship to infant-mother attachment. *Child Development* 41:291–312.

Belsky, J. 1976. Home and laboratory: the effect of setting on mother-infant interaction. Unpublished manuscript, Cornell University.

Belsky, J., and Steinberg, L. D. 1979. The effects of day care: a critical review. *Child Development,* in press.

Benedict, R. 1934. *Patterns of culture.* New York: Houghton Mifflin.

Beres, D., and Obers, S. 1950. The effects of extreme deprivation in infancy on psychic structure in adolescence. *Psychoanalytic Study of the Child* 5:212–235.

Bettelheim, B. 1943. Individual and mass behavior in extreme situations. *Journal of Abnormal and Social Psychology* 38:417–452.

Birenbaum, G. 1930. Das Vergessen einer Vornahme. *Psychologische Forschung* 13:218–284.

Bissell, J. S. 1971. Implementation of planned variation in Head Start: first year report. Washington, D.C.: Institute of Child Health and Human Development.

Blehar, M. 1974. Anxious attachment and defensive reactions associated with day care. *Child Development* 45:683–692.

Blenkner, M., Bloom, M., and Nielsen, M. 1971. A research and demonstration project of protective services. *Social Casework* 52:483–499.

Bott, E. 1957. *Family and social network: roles, norms, and external relationships in ordinary urban families.* London: Tavistock.

Bowlby, J. 1951. *Maternal care and mental health.* Geneva: World Health Organization.

Bowles, S., and Gintis, H. 1976. *Schooling in capitalist America: educational reform and the contradictions of economic life.* New York: Basic Books.

Bridges, K. M. B. 1932. Emotional development in early infancy. *Child Development* 3:324–341.

Bronfenbrenner, U. 1951. Toward an integrated theory of personality. In *Perception, an approach to personality,* ed. R. R. Black and G. V. Remsey. New York: Ronald Press.

Bronfenbrenner, U. 1961. Some familial antecedents of responsibility and leadership in adolescents. In *Leadership and interpersonal behavior,* ed. L. Petrullo and B. L. Bass. New York: Holt, Rinehart, and Winston.

Bronfenbrenner, U. 1967. Response to pressure from peers versus adults among Soviet and American school children. *International Journal of Psychology* 2:199–208.

Bronfenbrenner, U. 1968. Early deprivation: a cross-species analysis. In *Early experience and behavior,* ed. G. Newton and S. Levine. Springfield, Ill.: Charles C. Thomas.

Bronfenbrenner, U. 1970a. *Two worlds of childhood: U.S. and U.S.S.R.* New York: Russell Sage Foundation.

Bronfenbrenner, U. 1970b. Reaction to social pressure from adults versus peers among Soviet day school and boarding school pupils in the perspective of an American sample. *Journal of Personality and Social Psychology* 15:179–189.

Bronfenbrenner, U. 1974a. Developmental research, public policy, and the ecology of childhood. *Child Development* 45:1–5.

Bronfenbrenner, U. 1974b. The origins of alienation. *Scientific American* 231:53–61.

Bronfenbrenner, U. 1974c. Developmental research and public policy. In *Social science and social welfare,* ed. J. Romanyshyn. New York: Council on Social Work Education.

Bronfenbrenner, U. 1974d. *Is early intervention effective? A report on longitudinal evaluations of preschool programs,* vol. 2. Washington, D.C.: Department of Health, Education and Welfare, Office of Child Development.

Bronfenbrenner, U. 1975. Reality and research in the ecology of human development. *Proceedings of the American Philosophical Society* 119:439–469.

Bronfenbrenner, U. 1976. Research on the effects of day care and child development. In *Toward a national policy for children and families.* Washington, D.C.: National Academy of Sciences, Advisory Committee on Child Development.

Bronfenbrenner, U. 1977a. Toward an experimental ecology of human development. *American Psychologist* 32:513–531.

Bronfenbrenner, U. 1977b. Lewinian space and ecological substance. *Journal of Social Issues* 33:199–213.

Bronfenbrenner, U. 1978a. The social role of the child in ecological perspective. *Zeitschrift für Soziologie* 7:4–20.

Bronfenbrenner, U. 1978b. Who needs parent education? *Teachers College Record* 79:767–787.

Bronfenbrenner, U., Belsky, J., and Steinberg, L. 1976. Day care in context: an ecological perspective on research and public policy. Review prepared for the Office of the Assistant Secretary of Planning and Evaluation, Department of Health, Education and Welfare, Washington, D.C.

Bronfenbrenner, U., and Cochran, M. 1976. The comparative ecology of human development: a research proposal. Department of Human Development and Family Studies, Cornell University.

Brookhart, J., and Hock, E. 1976. The effects of experimental context and experiential background on infants' behavior toward their mothers and a stranger. *Child Development* 47:333–340.

Brunswik, E. 1943. Organismic achievement and environmental probability. *Psychological Review* 50:255–272.

Brunswik, E. 1956. Historical and thematic relations of psychology to other sciences. *Science Monitor* 83:151–161.

Brunswik, E. 1957. Scope and aspects of the cognitive problem. In *Contemporary approaches to cognition: a symposium held at the University of Colorado,* ed. H. Gruber, R. Jessor, and R. Hammond. Cambridge, Mass.: Harvard University Press.

Casler, L. 1961. Maternal deprivation: a critical review of the literature. *Monographs of the Society for Research in Child Development,* 26, no. 2.

Casler, L. 1968. Perceptual deprivation in institutional settings. In *Early experience and behavior,* ed. G. Newton and S. Levine. Springfield, Ill.: Charles C. Thomas.

Clarke, A. D. B., and Clarke, A. M. 1954. Cognitive changes in the feebleminded. *British Journal of Psychology* 45:173–179.

Clarke, A. D. B., and Clarke, A. M. 1959. Recovery from the effects of deprivation. *Acta Psychologica* 16:137–144.

Clarke, A. M., and Clarke, A. D. B. 1976. *Early experiences myth and evidence.* London: Open Books.

Clarke, A. D. B., Clarke, A. M., and Reiman, S. 1958. Cognitive and social changes in the feebleminded—three further studies. *British Journal of Psychology* 49:144–157.

Clausen, J. A. 1966. Family structure, socialization, and personality. In *Review of Child Development Research*, vol. 2, ed. L. W. Hoffman and M. L. Hoffman. New York: Russell Sage Foundation.

Cochran, M. M. 1973. A comparison of nursery and non-nursery child-rearing patterns in Sweden. Doctoral dissertation, University of Michigan.

Cochran, M. M. 1974. A study of group day care and family childrearing patterns in Sweden: second phase. Department of Human Development and Family Studies, Cornell University.

Cochran, M. M. 1975. The Swedish childrearing study: an example of the ecological approach to the study of human development. Paper presented at the Second Biennial Conference of the International Society for the Study of Behavioral Development, University of Surrey.

Cochran, M. M. 1977. A comparison of group day and family childrearing patterns in Sweden. *Child Development* 48:702–707.

Cochran, M. M., and Bronfenbrenner, U. 1978. Child rearing, parenthood, and the world of work. In *Work in America: the decade ahead,* ed. C. Kerr and J. M. Rosow. Scarsdale, N.Y.: Work in America Institute.

Cohen, S., Glass, D. C., and Singer, J. E. 1973. Apartment noise, auditory discrimination and reading ability in children. *Journal of Experimental Social Psychology* 9:407–422.

Cole, M., and Scribner, S., 1974. *Culture and thought: a psychological interpretation.* New York: John Wiley.

Cole, M., Gay, J., Glick, J. A., and Sharp, D. W. 1971. *The cultural context of learning and thinking.* New York: Basic Books.

Cole, M., Hood, L., and McDermott, R. P. 1978. Concepts of ecological validity: their differing implications for comparative cognitive research. *Quarterly Newsletter of the Institute for Comparative Human Development* 2:34–37.

Cole, M., and I. Maltzman, eds. 1969. *A handbook of contemporary Soviet psychology.* New York: Basic Books.

Committee on the Judiciary of the United States Senate. 1975. *Our nation's schools—a report card: "A" in school violence and vandalism.* Preliminary report of the Subcommittee to Investigate Juvenile Delinquency. Washington, D.C.: U.S. Government Printing Office.

Cooley, C. H. 1902. *Human nature and the social order.* New York: Scribner.

Cottrell, L. S. 1942. The analysis of situational fields in social psychology. *American Sociological Review* 7:370–382.

Cronbach, L. J., and Meehl, P. E. 1955. Construct validity in psychological tests. *Psychological Bulletin* 52:281–302.

DeJong, W. 1975. Another look at Banuazizi and Movahedi's analysis of the Stanford prison experiment. *American Psychologist* 30:1013–15.

Dennis, W. 1960. Causes of retardation among institutional children: Iran. *Journal of Genetic Psychology* 96:47–59.

Dennis, W., and Najarian, P. 1957. Infant development under environmental handicaps. *Psychological Monographs* 71, no. 7.

Dennis, W., and Sayegh, Y. 1965. The effect of supplementary experiences upon the behavioral development of infants in institutions. *Child Development* 36:81–90.

Department of Health, Education, and Welfare. 1968. *Federal Interagency Day Care Requirements*. DHEW Publication no. (OHD) 76-31081.

Devereux, E. C., Bronfenbrenner, U., and Rodgers, R. R. 1969. Child-rearing in England and the United States: a cross-national comparison. *Journal of Marriage and the Family* 31:257–270.

Devereux, E. C., Bronfenbrenner, U., and Suci, G. J. 1962. Patterns of parent behavior in America and West Germany: a cross-national comparison. *International Social Science Journal* 14:488–506.

Devereux, E. C., Shouval, R., Bronfenbrenner, U., Rodgers, R. R., Kav-Venaki, S., Kiely, E., and Karson, E. 1974. Socialization practices of parents, teachers, and peers in Israel: the kibbutz versus the city. *Child Development* 45:269–281.

Dewey, J. 1913. *The school in society*. Chicago: University of Chicago Press.

Dewey, J. 1916. *Democracy and education: an introduction to the philosophy of education*. New York: Macmillan.

Dewey, J. 1931. *The child and the curriculum*. Chicago: University of Chicago Press.

DiLorenzo, L. T. 1969. Pre-kindergarten programs for educationally disadvantaged children: final report. Washington, D.C.: U.S. Office of Education.

Doll, E. A. 1953. *The measurement of social competence: a manual for the Vineland social maturity scale*. Minneapolis: Educational Test Bureau, Educational Publishers.

Doyle, C. L. 1975a. Interpersonal dynamics in role playing. *American Psychologist* 30:1011–013.

Doyle, A. 1975b. Infant development in day care. *Developmental Psychology* 11:655–656.

Doyle, A., and Somers, K. The effects of group and family day care on infant attachment. Unpublished manuscript, Department of Psychology, Concordia University, Montreal, no date.

Elder, G. H., Jr. 1974. *Children of the Great Depression*. Chicago: University of Chicago Press.

Elder, G. H., Jr. 1979. Historical change in life pattern and personality. In *Lifespan development and behavior*, vol. 2, ed. P. Baltes and O. Brim. New York: Academic Press, in press.

Elder, G. H., Jr., and Rockwell, R. C. 1978. Economic depression and post-war opportunity: a study of life patterns in hell. In *Research in com-*

munity and mental health, ed. R. A. Simmons. Greenwich, Conn.: J A I Press, in press.

Elkonin, D. B. 1978. *Psikhologiya igry* (The psychology of play). Moscow: U.S.S.R. Academy of Pedagogical Sciences.

Elliot, V. 1973. Impact of day care on the economic status of the family. In *A summary of the Pennsylvania day care study,* ed. D. Peters. University Park: Pennsylvania State University.

Felner, R. D., Stolberg, A., and Cowan, E. L. 1975. Crisis events and school mental health referral patterns of young children. *Journal of Consulting Clinical Psychology* 43:305–310.

Fraiberg, S. H. 1977. *Every child's birthright: in defense of mothering.* New York: Basic Books.

Freud, S. 1933. *New introductory lectures on psycho-analysis.* New York: W. W. Norton.

Furstenberg, F. 1976. *Unplanned parenthood: the social consequences of teenage child bearing.* New York: Free Press.

Garbarino, J. 1975. A note on the effects of television viewing. In *Influences on human development,* 2d ed., ed. U. Bronfenbrenner and M. A. Mahoney. Hinsdale, Ill.: Dryden Press.

Garbarino, J. 1976. A preliminary study of some ecological correlates of child abuse: the impact of socioeconomic stress on mothers. *Child Development* 47:178–185.

Garbarino, J., and Bronfenbrenner, U. 1976. The socialization of moral judgment and behavior in cross-cultural perspective. In *Moral development and behavior,* ed. T. Lickona. New York: Holt, Rinehart, and Winston.

Geismar, L. L., and Ayres, B. 1960. Measuring family functioning. St. Paul, Minn.: Family Center Project, St. Paul United Fund and Council.

Gesell, A. L. 1948. *The first five years of life.* New York: Harper.

Getzels, J. W. 1969. A social psychology of education. In *The handbook of social psychology,* 2d ed., vol. 5, ed. G. Lindzey and E. Aronson. Reading, Mass.: Addison-Wesley.

Gewirtz, J. L. 1965. The course of infant smiling in four child-rearing environments in Israel. In *Determinants of infant behaviour,* vol. 3, ed. B. M. Foss. New York: John Wiley.

Gilmer, B., Miller, J. O., and Gray, S. W. 1970. Intervention with mothers and young children: study of intra-family effects. Nashville, Tenn.: DARCEE Demonstration and Research Center for Early Education.

Giovannoni, J., and Billingsley, A. 1970. Child neglect among the poor: a study of parental adequacy in families of three ethnic groups. *Child Welfare* 49:196–204.

Glick, P. C. 1978. Social change in the American family. *Social Welfare Forum, 1977.* New York: Columbia University Press.

Golden, M., Rosenbluth, L., Grossi, M., Policare, H., Freeman, H., and

Brownlee, E. 1978. The New York City infant day care study. New York: Medical and Health Research Association of New York City.

Goldfarb, W. 1943a. The effects of early institutional care on adolescent personality. *Journal of Experimental Education* 12:106–129.

Goldfarb, W. 1943b. Infant rearing and problem behavior. *American Journal of Orthopsychiatry* 13:249–265.

Goldfarb, W. 1955. Emotion and intellectual consequences of psychological deprivation in infancy: a re-evaluation. In *Psychopathology of childhood,* ed. P. H. Hoch and J. Zubin. New York: Grune and Stratton.

Graves, Z. R., and Glick, J. 1978. The effect of context on mother-child interaction: a progress report. *The Quarterly Newsletter of the Institute for Comparative Human Development* 2:41–46.

Gunnarsson, L. 1973. Family day care homes: an alternative form of child care. Masters thesis, University of Michigan.

Gunnarrsson, L. 1978. Children in day care and family care in Sweden: a follow-up. Doctoral dissertation, University of Michigan.

Hales, D. 1977. How early is early contact? Defining the limits of the sensitive period. Paper prepared for presentation at the meeting of the Society for Research in Child Development, New Orleans.

Hales, D., Kennell, J. H., and Susa, R. 1976. How early is early contact? Defining the limits of the sensitive period. Report to the Foundation for Child Development on the Ecology of Human Development Program. New York: Foundation for Child Development.

Haney, C., Banks, C., and Zimbardo, P. 1973. Interpersonal dynamics in a simulated prison. *International Journal of Criminology and Penology* 1:69–97.

Harnischfeger, A., and Wiley, D. E. 1975. *Achievement test score decline: do we need to worry?* Chicago: *CEMREL.*

Harrell, J. 1973. Substitute child care, maternal employment and the quality of mother-child interaction. In A summary of the Pennsylvania day care study, ed. D. Peters. Pennsylvania State University, mimeographed.

Harrell, J., and Ridley, C. 1975. Substitute child care, maternal employment and the quality of mother-child interaction. *Journal of Marriage and the Family* 37:556–565.

Hartup, W. W. 1970. Peer interaction and social organization. In *Manual of child psychology,* vol. 2, ed. P. H. Mussen. New York: John Wiley.

Hayes, D., and Grether, J. 1969. The school year and vacation: when do students learn? Paper presented at the Eastern Sociological Convention, New York.

Heinicke, C. 1956. Some effects of separating two-year-old children from their parents: a comparative study. *Human Relations* 9:105–176.

Heinicke, C. M., and Westheimer, I. J. 1965. *Brief separations.* New York: International Universities Press.

Hertzig, M. E., Birch, H. G., Thomas, A., and Mendez, O. A. 1968. Class and ethnic differences in the responsiveness of preschool children to cognitive demands. *Monographs of the Society for Research in Child Development* 33:no. 1, serial no. 117.

Hess, R. D. 1970. Social class and ethnic influences on socialization. In *Manual of child psychology*, vol. 2, ed. P. H. Mussen. New York: John Wiley.

Hetherington, E. M. 1972. Effect of paternal absence on personality development in adolescent daughters. *Developmental Psychology* 7: 313–326.

Hetherington, E. M., Cox, M., and Cox, R. 1976. Divorced fathers. *The Family Coordinator* 25:417–428.

Hetherington, E. M., Cox, M., and Cox, R. 1977. The development of children in mother-headed families. Paper presented at the Conference of Families in Contemporary America. Washington, D.C.: George Washington University.

Hetherington, E. M., Cox, M., and Cox, R. 1978. The aftermath of divorce. In *Mother-child, father-child relations*, ed. J. H. Stevens and M. Mathews. Washington, D.C.: National Association for the Education of Young Children.

Heyns, B. (no date) *Exposure and the effects of schooling.* Cambridge, Mass. Center for the Study of Public Policy.

Husserl, E. 1950. *Ideen zu einer reinen Phänomenologie und phänomenologischen Philosophie.* Haag: M. Nijhoff.

Jones, N. B., ed. 1972. *Ethological studies of child behavior.* Cambridge: Cambridge University Press.

Kagan, J., Kearsley, R., and Zelazo, P. 1978. *Infancy: its place in human development.* Cambridge, Mass.: Harvard University Press.

Karnes, M. B. 1969. Research and development program on preschool disadvantaged children: final report. Washington, D.C.: U.S. Office of Education.

Karnes, M. B., Hodgins, A. S., and Teska, J. A. 1969. The impact of at-home instruction by mothers on performance in the Ameliorative preschool. In Research and development program on preschool disadvantaged children: final report, ed. M. B. Karnes. Washington, D.C.: U.S. Office of Education.

Karnes, M. B., Teska, J. A., Hodgins, A. S., and Badger, E. D. 1970. Educational intervention at home by mothers of disadvantaged infants. *Child Development* 41:925–935.

Katz, D. 1911. *Die Erscheinungsweisen der Farben.* Leipzig: Barth.

Katz, D. 1930. *Der Aufbau der Farbwelt.* Leipzig: Barth.

Kav-Venaki, S., Eyal, N., Bronfenbrenner, U., Kiely, E., and Caplan, D. 1976. The effect of Russian versus Hebrew instructions on the reaction to social pressure of Russian-born Israeli children. *Journal of Experimental Social Psychology* 12:70–86.

Kennell, J. H., Jerauld, R., Wolfe, H., Chesler, D., Kreger, N. C., McAlpine, W., Steffa, J., and Klaus, M. H. 1974. Maternal behavior one year after early and extended post-partum contact. *Developmental Medicine and Child Neurology* 16:172–179.

Kennell, J. H., Trause, M. A., and Klaus, M H. 1975. Evidence for a sensitive period in the human mother. In *CIBA Foundation Symposium 33* (new series). Amsterdam: Elsevier.

Kessen, W., ed. 1975. *Childhood in China*. New Haven: Yale University Press.

Klaus, M. H., and Kennell, J. H. 1976. *Maternal-infant bonding*. St. Louis: Mosby.

Klaus, M. H., Jerauld, R., Kreger, N. C., McAlpine, W., Steffa, M., and Kennell, J. H. 1972. Maternal attachment: importance of the first post-partum day. *New England Journal of Medicine* 286:460–463.

Klaus, M. H., Kennell, J. H., Plumb, N., and Zuehlke, S. 1970. Human maternal behavior at the first contact with her young. *Pediatrics* 46: 187–192.

Klaus, M. H., Trause, M. A., and Kennell, J. H. 1975. Does human maternal behavior after a delivery show a characteristic pattern? In *CIBA Foundation Symposium 33* (new series). Amsterdam: Elsevier.

Koffka, K. 1935. *Principles of Gestalt psychology*. New York: Harcourt Brace.

Kogan, K., and Wimberger, H. 1969. Interaction patterns in disadvantaged families. *Journal of Clinical Psychology* 25:347–352.

Köhler, K. 1929. *Gestalt psychology*. New York: Liveright.

Köhler, K. 1938. *The place of value in a world of facts*. New York: Liveright.

Kohn, M. L. 1963. Social class and parent-child relationships: an interpretation. *American Journal of Sociology* 68:471–480.

Kohn, M. L. 1969. *Class and conformity: a study in values*. Homewood, Ill.: Dorsey Press.

Kohn, M. L. 1977. Class and conformity: reassessment, 1977. In *Class and conformity: a study in values*, 2d ed., ed. M. Kohn. Chicago: University of Chicago Press.

Kohn, M. L., and Schooler, C. 1973. Occupational experience and psychological functioning: an assessment of reciprocal effects. *American Sociological Review* 38:97–118.

Kohn, M. L., and Schooler, C. 1978. The reciprocal effects of the substantive complexity of work and intellectual flexibility: a longitudinal assessment. *American Journal of Sociology* 84:24–52.

Kotelchuck, M., Zelazo, P., Kagan, J., and Spelke, E. 1975. Infant reaction to parental separations when left with familiar and unfamiliar adults. *Journal of Genetic Psychology* 126:255–262.

Labov, W. 1967. Some sources of reading problems for Negro speakers of non-standard English. In *New directions in elementary English,* ed. A. Frazier. Champaign, Ill.: National Council of Teachers of English.

Labov, W. 1970. The logic of nonstandard English. In *Language and poverty*, ed. F. Williams. Chicago: Markham.

Lally, R. 1973. The family development research program, progress report. Syracuse University.

Lally, R. 1974. The family development research program, progress report. Syracuse University.

Lamb, M. E. 1975. Infants, fathers, and mothers: interaction at 8 months of age in the home and in the laboratory. Paper presented at the meeting of the Eastern Psychological Association, New York.

Lamb, M. E. 1976a. The role of the father: an overview. In *The role of the father in child development*, ed. M. E. Lamb. New York: John Wiley.

Lamb, M. E. 1976b. Effects of stress and cohort on mother- and father-infant interaction. *Developmental Psychology* 12:435–443.

Lamb, M. E. 1976c. Interactions between eight-month-old children and their fathers and mothers. In *The role of the father in child development*, ed. M. E. Lamb. New York: John Wiley.

Lamb, M. E. 1977. The development of mother-infant and father-infant attachments in the second year of life. *Developmental Psychology* 13:637–648.

Lamb, M. E. 1978. Infant social cognition and small "second order" effects. *Infant Behavior and Development* 1:1-10.

Larson, M. T. 1975. Current trends in child development research. Unpublished manuscript, School of Home Economics, University of North Carolina.

Lay, M., and Meyer, W. 1973. *Teacher/child behaviors in an open environment day care program.* Syracuse University Children's Center, mimeographed.

Lazar, I., and Darlington, R. B. 1978. Lasting effects after preschool. New York State College of Human Ecology, Cornell University.

Lazar, I., Hubbell, B. R., Murray, H., Rosche, M., and Royce, J. 1977a. Persistence of preschool effects: final report. Grant no. 18–76–07843 to the Administration on Children, Youth and Families. Washington, D.C.: Office of Human Development Services, U.S. Department of Health, Education and Welfare.

Lazar, I., Hubbell, B. R., Murray, H., Rosche, M., and Royce, J. 1977b. Summary: the persistence of preschool effects; summary of final report. Grant no. 18–76–07843 to the Administration on Children, Youth and Families. Washington, D.C.: Office of Human Development Services, U.S. Department of Health, Education and Welfare.

Leontiev, A. N. 1964. *Problems of mental development.* Washington, D.C.: U.S. Joint Publications Research Service.

Lester, B. M., Kotelchuck, M., Spelke, E., Sellers, M. J., and Klein, R. E.

1974. Separation protest in Guatemalan infants: cross-cultural and cognitive findings. *Developmental Psychology* 10:79–85.

Levenstein, P. 1970. Cognitive growth in preschoolers through verbal interaction with mothers. *American Journal of Orthopsychiatry* 40: 426–432.

Lewin, K. 1917. Kriegslandschaft. *Zeitschrift für Angewandte Psychologie* 12:440–447.

Lewin, K. 1931. Environmental forces in child behavior and development. In *A handbook of child psychology*, ed. C. Murchison. Worcester, Mass.: Clark University Press.

Lewin, K. 1935. *A dynamic theory of personality.* New York: McGraw-Hill.

Lewin, K. 1943. Defining the "field at a given time." *Psychological Review* 50:292–310.

Lewin, K. 1948. *Resolving social conflicts, selected papers on group dynamics.* New York: Harper.

Lewin, K. 1951. *Field theory in social science, selected theoretical papers.* New York: Harper.

Lewin, K., Lippitt, R., and White, R. K. 1939. Patterns of aggressive behavior in experimentally created "social climates." *Journal of Social Psychology* 10:271–299.

Liebert, R. M., Neale, J. M., and Davidson, E. S. 1973. *The early window: effects of television on children and youth.* New York: Pergamon Press.

Linton, R. 1936. *The study of man.* New York: Appleton-Century.

Lippitt, R. 1940. An experimental study of the effect of democratic and authoritarian group atmospheres. *Studies in topological and vector psychology I.* University of Iowa Studies in Child Welfare 16: 44–195.

Lippman, M. A., and Grote, B. H. 1974. Socio-emotional effects of day care: final project report. Bellingham, Wash.: Western Washington State College.

Littenberg, R., Tulkin, S. R., and Kakan, J. 1971. Cognitive components of separation anxiety. *Developmental Psychology* 4:387–388.

Luria, A. R. 1976. *Cognitive development: its cultural and social foundations.* Cambridge, Mass.: Harvard University Press.

Lüscher, K. 1971. Dreizehnjährige Schweizer zwischen Peers und Erwachsenen im interkulturellen Vergleich. (13-year-old Swiss children between peers and adults in cross-cultural comparison.) *Schweizerische Zeitschrift für Psychologie und Ihre Andwendungen* 30:219–229.

Lüscher, K., and Fisch, R. 1977. Das Sozialisationswissen junger Eltern (The socialization knowledge of young parents). Konstanz, West Germany: University of Konstanz, Project Group "Familiäre Sozialisation."

Maas, H. 1963. Long-term effects of early childhood separation and group care. *Vita Humana* 6:34–56.

Maccoby, E. E. 1951. Television: its impact on school children. *Public Opinion Quarterly* 15:421–444.

MacLeod, R. B. 1947. The phenomenological approach to social psychology. *Psychological Review* 54:193–210.

Macrae, J. W., and Herbert-Jackson, E. 1976. Are behavioral effects of infant day care program specific? *Developmental Psychology* 12:269–270.

Mayhew, H. 1968. *London labour and the London poor*, vol. 1. New York: Dover Publications (originally published by Griffin, Bohn, in 1861–62.)

McAllister, R. J., Butler, E. W., and Lei, T. 1973. Patterns of social interaction among families of behaviorally retarded children. *Journal of Marriage and the Family* 35:93–100.

McCall, R. B. 1977. Challenges to a science of developmental psychology. *Child Development* 48:333–344.

McCutcheon, B., and Calhoun, K. 1976. Social and emotional adjustment of infants and toddlers to a day care setting. *American Journal of Orthopsychiatry* 46:104–108.

McGrew, W. C. 1972. *An ethological study of children's behavior*. New York: Academic Press.

McNeil, E. B. 1962. Waging experimental war: a review. *Journal of Conflict Resolution* 6:77–81.

Mead, G. H. 1934. *Mind, self, and society*. Chicago: University of Chicago Press.

Mercer, J. 1971. Sociocultural factors in labeling mental retardates. *The Peabody Journal of Education* 48:188–203.

Meyers, L. 1973. The relationship between substitute child care, maternal employment and female marital satisfaction. In A summary of the Pennsylvania day care study, ed. D. Peters. Pennsylvania State University, mimeographed.

Milgram, S. 1963. Behavioral study of obedience. *Journal of Abnormal and Social Psychology* 67:371–378.

Milgram, S. 1964. Group pressure and action against a person. *Journal of Abnormal and Social Psychology* 69:137–143.

Milgram, S. 1965a. Liberating effects of group pressure. *Journal of Personality and Social Psychology* 1:127–134.

Milgram, S. 1965b. Some conditions of obedience and disobedience to authority. *Human Relations* 18:57–76.

Milgram, S. 1974. *Obedience to authority*. New York: Harper & Row.

Mills, C. W. 1956. *The power elite*. New York: Oxford University Press.

Mitchell, J. C. ed. 1969. *Social networks in urban situations*. Manchester: Manchester University Press.

Moore, T. 1964 Children of full-time and part-time mothers. *International Journal of Social Psychiatry,* Special Congress Issue #2:1–10.

Moore, T. 1972. The later outcomes of early care by the mother and substitute daily regimes. In *Determinants of behavioral development,* ed. F. J. Monks, W. W. Hartup, and J. deWitt. New York: Academic Press.

Moore, T. 1975. Exclusive early mothering and its alternatives: the outcome to adolescence. *Scandinavian Journal of Psychology* 16:255–272.

Morgan, G. A., and Ricciuti, H. N. 1965. Infants' responses to strangers during the first year. In *Determinants of infant behavior,* vol. 4, ed. B. M. Foss. London: Methuen.

Moskowitz, D. S., Schwarz, J. C., and Corsini, D. A. 1977. Initiating day care at three years of age: effects on attachment. *Child Development* 48:1271–76.

Moustakas, C. E., Sigel, I. E., and Schalock, H. D. 1956. An objective method for measurement and analysis of child-adult behavior interaction. *Child Development* 27:109–134.

Nerlove, S., Bronfenbrenner, U., Blum, K., Robinson, J., and Koel, A. 1978. Transcultural code of molar activities of children and caretakers in modern industrialized societies. Department of Human Development and Family Studies, Cornell University, mimeographed.

O'Connor, N. 1956. The evidence for the permanently disturbing effects of mother-child separation. *Acta Psychologica* 12:174–191.

O'Connor, N. 1968. Children in restricted environments. In *Early experience and behavior,* ed. G. Newton and S. Levine. Springfield, Ill.: Charles C. Thomas.

Ogbu, J. U. 1974. *The next generation: an ethnography of education in an urban neighborhood.* New York: Academic Press.

Orlansky, H. 1949. Infant care and personality. *Psychological Bulletin* 46:1–48.

Orne, M. T. 1962. On the social psychology of the experiment: with particular reference to demand characteristics and their implications. *American Psychologist* 17:776–783.

Orne, M. T. 1973. Communication by the total experimental situation: why is it important, how it is evaluated, and its significance for the ecological validity of findings. In *Communication and affect,* ed. P. Pliner, L. Krames, and T. Alloway. New York: Academic Press.

O'Rourke, J. F. 1963. Field and laboratory: the decision-making behavior of family groups in two experimental conditions. *Sociometry* 26:422–435.

Ovsiankina, M. 1928. Die Wiederaufnahme unterbrochener Handlungen. (The resumption of interrupted actions.) *Psychologische Forschung* 11:302–379.

Parke, R. D. 1978. Parent-infant interaction: progress, paradigms, and problems. In *Observing behavior, vol. 1, theory and application in mental retardation,* ed. G. P. Sackett. Baltimore: University Park Press.

Parke, R. D. 1979. Interactional design and experimental manipulation: the field lab interface. In *Social interaction: methods, analysis and illustration,* ed. R. B. Cairns. Hillsdale, N.J.: Erlbaum Associates, in press.

Parsons, T. 1955. Family structure and the socialization of the child. In *Family, socialization and interaction process,* ed. T. Parsons and R. F. Bales. Glencoe, Ill.: Free Press.

Parsons, T., and Bales, R. F. 1955. *Family, socialization, and interaction process.* Glencoe, Ill.: Free Press.

Pederson, F. A. 1976. Mother, father, and infant as an interaction system. Paper presented at the annual meeting of the American Psychological Association, Washington, D.C.

Piaget, J. 1954. *The construction of reality in the child.* New York: Basic Books.

Piaget, J. 1962. *Play, dreams and imitation in childhood.* London: Routledge and Paul.

Piliavin, I. M., Rodin, J., and Piliavin, J. A. 1969. Good samaritanism: an underground phenomenon? *Journal of Personality and Social Psychology* 13:289–299.

Pinneau, S. 1955. The infantile disorders of hospitalism and anaclitic depression. *Psychological Bulletin* 52:429–452.

Portnoy, F., and Simmons, C. 1978. Day care and attachment. *Child Development* 49:239–242.

Prescott, E. 1973. A comparison of three types of day care and nursery school-home care. Paper presented at the Society for Research in Child Development, Philadelphia.

Pringle, M. L., and Bossio, B. 1958. A study of deprived children. *Vita Humana* 1:65–92, 142–170.

Provence, S., and Lipton, R. 1962. *Infants in institutions.* New York: International University Press.

Prugh, D. G., Staub, E. M., Sands, H. H., Kirschbaum, R. M., and Lenihan, E. A. 1953. A study of the emotional reactions of children in families to hospitalization and illness. *American Journal of Orthopsychiatry* 23:70–106.

Ramey, C., and Campbell, F. 1977. The prevention of developmental retardation in high-risk children. In *Research to practice in mental retardation, vol. 1, Care and intervention,* ed. P. Mittler. Baltimore: University Park Press.

Ramey, C., and Smith, B. 1977. Assessing the intellectual consequences of early intervention with high-risk infants. *American Journal of Mental Deficiency* 81:318–324.

Raph, J. B., Thomas, A., Chess, S., and Korn, S. J. 1968. The influence of nursery school on social interactions. *American Journal of Ortho-psychiatry* 38:144–152.

Razmyslov, P. 1934. Vygotsky and Luria's cultural-historical theory of psychology. *Moscow: Knigii Proletarskoi Revolutsii* 4:78–86.

Rheingold, H. L., ed. 1963. *Maternal behavior in mammals.* New York: John Wiley.

Rheingold, H. L. 1969a. The social and socializing infant. In *Handbook of socialization theory and research,* ed. D. A. Goslin. Chicago: Rand McNally.

Rheingold, H. L. 1969b. The effect of a strange environment on the behavior of infants. In *Determinants of infant behavior,* vol. 4, ed. B. M. Foss. London: Methuen.

Rheingold, H. L., and Eckerman, C. O. 1970. The infant separates himself from his mother. *Science* 168:78–90.

Ricciuti, H. N. 1974. Fear and development of social attachments in the first year of life. In *The origins of human behavior: fear,* eds. M. Lewis and L. A. Rosenblum. New York: John Wiley.

Ricciuti, H. N. 1976. Effects of infant day care experience on behavior and development: research and implications for social policy. Paper prepared for the Office of the Assistant Secretary for Planning and Evaluation. Department of Health, Education, and Welfare.

Ringler, N. 1977. Mothers' speech to her two-year-old child: its effect on speech and language comprehension at five years. Paper presented at the Annual Meeting of the Pediatric Research Society, St. Louis.

Ringler, H., Kennell, J. H., Jarvella, R., Navojosky, R. J., and Klaus, M. H. 1975. Mother-to-child speech at two years—effects of early postnatal contact. *Journal of Pediatrics* 86:141–144.

Rodgers, R. R. 1971. Changes in parental behavior reported by children in West Germany and the United States. *Human Development* 14: 208–224.

Rodgers, R. R., Bronfenbrenner, U., and Devereux, E. C. 1968. Standards of social behavior among children in four cultures. *International Journal of Psychology* 3:31–41.

Rosenthal, R., and Jacobson, L. 1968. *Pygmalion in the classroom: teacher expectation and pupils' intellectual development.* New York: Holt, Rinehart, and Winston.

Ross, G., Kagan, J., Zelazo, P., and Kotelchuck, M. 1975. Separation protest in infants in home and laboratory. *Developmental Psychology* 11:256–257.

Santrock, J. W. 1975. Father absence, perceived maternal behavior, and moral development in boys. *Child Development* 46:753–757.

Sarbin, T. R. 1968. Role: psychological aspects. In *International encyclopedia of the social sciences,* vol. 13. New York: Macmillan.

Scarr-Salapatek, S., and Williams, M. L. 1973. The effects of early stimulation on low-birth weight infants. *Child Development* 44:94–101.

Schaefer, E. S. 1968. Progress report: intellectual stimulation of culturally-deprived parents. National Institute of Mental Health.

Schaefer, E. S. 1970. Need for early and continuing education. In *Education of the infant and young child*, ed. V. H. Denenberg. New York: Academic Press.

Schaefer, E. S., and Aaronson, M. 1972. Infant education research project: implementation and implications of the home-tutoring program. In *The preschool in action*, ed. R. K. Parker. Boston: Allyn and Bacon.

Schaffer, H. R. 1958. Objective observations of personality development in early infancy. *British Journal of Medical Psychology* 31:174–183.

Schaffer, H. R. 1963. Some issues for research in the study of attachment behavior. In *Determinants of infant behavior*, vol. 2, ed. B. M. Foss. New York: John Wiley.

Schaffer, H. R. 1965. Changes in developmental quotient under two conditions of maternal separation. *British Journal of Social and Clinical Psychology* 4:39–46.

Schaffer, H. R., and Callender, W. M. 1959. Psychologic effects of hospitalization in infancy. *Pediatrics* 24:528–539.

Schaffer, H. R., and Emerson, P. E. 1964. The development of social attachments in infancy. *Monographs of the Society for Research in Child Development* 29;no. 3, serial no. 94.

Schalock, H. D. 1956. Observation of mother-child interaction in the laboratory and in the home. *Dissertation Abstracts* 16:707.

Schlieper, A. 1975. Mother-child interaction at home. *American Journal of Orthopsychiatry* 45:468–472.

Schwarz, J. C., and Wynn, R. 1971. The effects of mothers' presence and previsits on children's emotional reaction to starting nursery school. *Child Development* 42:871–881.

Schwarz, J. C., Krolick, G., and Strickland, R. G. 1973. Effects of early day care experience on adjustment to a new environment. *American Journal of Orthopsychiatry* 43:340–346.

Schwarz, J. C., Strickland, R. G., and Krolick, G. 1974. Infant day care: behavioral effects at preschool age. *Developmental Psychology* 10:502–506.

Seaver, W. B. 1973. Effects of naturally induced teacher expectancies. *Journal of Personality and Social Psychology* 28:333–342.

Seitz, V., Abelson, W. D., Levine, E., and Zigler, E. 1975. Effects of place of testing on the Peabody Picture Vocabulary Test scores of disadvantaged Head Start and non-Head Start children. *Child Development* 46:481–486.

Shapira, A., and Madsen, M. C. 1969. Cooperative and competitive behavior of kibbutz and urban children in Israel. *Child Development* 40:609–617.

Sherif, M. 1956. Experiments in group conflicts. *Scientific American* 195:54–58.

Sherif, M., Harvey, O. J., Hoyt, B. J., Hood, W. R., and Sherif, C. W. 1961. *Intergroup conflict and cooperation: the robbers cave experiment*. Norman: University of Oklahoma Book Exchange.

Shouval, R. H., Kav-Venaki, S., Bronfenbrenner, U., Devereux, E. C., and Kiely, E. 1975. Anomalous reactions to social pressure of Israeli and Soviet children raised in family versus collective settings. *Journal of Personality and Social Psychology* 32:477–489.

Skeels, H. M. 1966. Adult status of children with contrasting early life experience. *Monographs of the Society for Research in Child Development* 31:no. 3, serial no. 105.

Skeels, H. M., and Dye, H. B. 1939. The study of the effects of differential stimulation on mentally retarded children. *Proceedings and Addresses of the American Association of Mental Deficiency* 44: 114–136.

Skeels, H. M., Updegraff, R., Wellman, B. L., and Williams, H. M. 1938. A study of environmental stimulation: an orphanage preschool project. *University of Iowa Studies in Child Welfare* 15:no. 4.

Smith, M. B. 1968. School and home: focus on achievement. In *Developing programs for the educationally disadvantaged*, ed. A. H. Passow. New York: Teachers College Press.

Soar, R. S. 1966. An integrative approach to classroom learning. NIMH Project No. 5-R11MH01096 to the University of South Carolina and 7-R11MH02045 to Temple University.

Soar, R. S. 1972. Follow-Through classroom process measurement and pupil growth (1970–71). College of Education, University of Florida, mimeographed.

Soar, R. S., and Soar, R. M. 1969. Pupil subject matter growth during summer vacation. *Educational Leadership Research Supplement* 26: 577–587.

Spelke, E., Zelazo, P., Kagan, J., and Kotelchuck, M. 1973. Father interaction and separation protest. *Developmental Psychology* 9:83–90.

Spitz, R. A. 1945. Hospitalism: an inquiry into the genesis of psychiatric conditions in early childhood. *Psychoanalytic Study of the Child* 1:153–172.

Spitz, R. A. 1946a. Hospitalism: a follow-up report on investigation described in volume 1, 1945. *Psychoanalytic Study of the Child* 2:113–117.

Spitz, R. A. 1946b. Anaclitic depression: an inquiry into the genesis of psychiatric conditions in early childhood, II. *Psychoanalytic Study of the Child* 2:313–342.

Spitz, R. A. 1946c. The smiling response: a contribution to the ontogenesis of social relations. *Genetic Psychology Monographs* 34: 57–125.

Sroufe, L. A. 1970. A methodological and philosophical critique of intervention-oriented research. *Developmental Psychology* 2:140–145.

Sroufe, L. A., Waters, E., and Matas, L. 1974. Contextual determinants of infant affective response. In *The origins of fear,* ed. M. Lewis, and L. A. Rosenblum. New York: John Wiley.

Stanford Research Institute, 1971a. Implementation of planned variation in Head Start: preliminary evaluation of planned variation in Head Start according to Follow-Through approaches (1969–70). Washington, D.C.: Office of Child Development, U.S. Department of Health, Education and Welfare.

Stanford Research Institute, 1971b. Longitudinal evaluation, selected features of the national Follow-Through program. Washington, D.C.: U.S. Department of Health, Education and Welfare.

Sullivan, H. S. 1947. *Conceptions of modern psychiatry.* Washington, D.C.: William Alanson White Psychiatric Foundation.

Thayer, S., and Saarni, C. 1975. Demand characteristics are everywhere (anyway): a comment on the Stanford prison experiment. *American Psychologist* 30:1015–16.

Thomas, W. I. 1927. *The unadjusted girl.* Boston: Little, Brown.

Thomas, W. I., and Thomas, D. S. 1928. *The child in America.* New York: Alfred P. Knopf.
York: Alfred A. Knopf.

Thomas, W. I., and Znaniecki, F. 1927. *The Polish peasant in Europe and America.* New York: Alfred A. Knopf.

Tizard, B., Cooperman, O., Joseph, A., and Tizard, J. 1972. Environmental effects on language development: a study of young children in long-stay residential nurseries. *Child Development* 43:337–358.

Tizard, B., and Hodges, J. 1978. The effect of early institutional rearing on the development of eight year old children. *Journal of Child Psychology and Psychiatry* 19:99–118.

Tizard, B., and Rees, J. 1974. A comparison of the effects of adoption, restoration to the natural mother, and continued institutionalization on the cognitive development of four-year-old children. *Child Development* 45:92–99.

Tizard, B., and Rees, J. 1976. A comparison of the effects of adoption, restoration to the natural mother, and continued institutionalization on the cognitive development of four-year-old children: further note: December 1975. In *Early experience: myth and evidence,* ed. A. M. Clarke and A. D. B. Clarke. London: Open Books.

Travers, J., and Ruopp, R. 1978. National day care study: preliminary findings and their implications. Cambridge, Mass., Abt Associates, mimeographed.

Tuckman, J., and Regan, R. A. 1966. Intactness of the home and behavioral problems in children. *Journal of Child Psychology and Psychiatry* 7:225–234.

Tulkin, S. S. 1972. An analysis of the concept of cultural deprivation. *Developmental Psychology* 6:326–339.

U.S. Bureau of the Census 1977. Money, income, and poverty status of families and persons in the United States: 1976 (Advance report). Current population reports, series P-60, no. 107. Washington, D.C.: U.S. Government Printing Office.

U.S. Bureau of the Census 1978. Marital status and living arrangements: March 1977. Current population reports, series P-20, no. 323. Washington, D.C.: U.S. Government Printing Office.

Venger, L. A. 1973. *Pedagogika sposobnostei* (The education of abilities). Moscow: Academy of Pedagogical Sciences.

Vopava, J., and Royce, J. 1978. Comparison of the long term effects of infant and preschool programs on academic performance. Paper presented as part of a symposium on early intervention programs at the Annual Meeting of the American Educational Research Association, Toronto, March 27.

Vygotsky, L. S. 1962. *Thought and language.* Cambridge, Mass.: M.I.T. Press.

Vygotsky, L. S. 1978. *Mind in society: the development of higher psychological processes.* Cambridge, Mass.: Harvard University Press.

Walters, J., Connor, R., and Zunich, M. 1964. Interaction of mothers and children from lower-class families. *Child Development* 35:433–440.

Weinraub, M. 1977. Children's responses to maternal absence: an experimental intervention study. Paper presented at the Society for Research in Child Development meetings, New Orleans.

Weinraub, M., and Lewis, M. 1977. The determinants of children's responses to separation. *Monographs of the Society for Research in Child Development* 42:no. 4., serial no. 172.

Weisz, J. 1978. Transcontextual validity in developmental research. *Child Development* 49:1–12.

Wertheimer, M. 1912. Experimentelle Studien über das Sehen von Bewegung. *Zeitschrift für Psychologie* 61:161–265.

White, R. K., and Lippitt, R. 1960. *Autocracy and democracy: an experimental inquiry.* New York: Harper.

Wright, H. F. 1967. *Recording and analyzing child behavior.* New York: Harper & Row.

Yarrow, L. J. 1956. The development of object relationships during infancy and the effects of a disruption of early mother-child relationships. *American Psychologist* 11:423 (abstract).

Yarrow, L. J. 1961. Maternal deprivation: toward an empirical and conceptual re-evaluation. *Psychological Bulletin* 58:459–490.

Yarrow, L. J. 1964. Separation from parents during early childhood. In *Review of child development research,* vol. 1, ed. M. L. Hoffman and L. Hoffman. New York: Russell Sage Foundation.

Yarrow, L. J., and Goodwin, M. S. 1963. Effects of change in mother figure during infancy on personality development. Progress report, Family and Child Services, Washington, D.C.

Zaporozhets, A. V., and Elkonin, D. B. 1971. *The psychology of preschool children.* Cambridge, Mass.: M.I.T. Press.

Zaporozhets, A. V., and Markova, T. A. 1976. *Vospitanie i obuchenie v detskom sadu* (Upbringing and instruction in kindergarten). Moscow: Pedagogika.

Zelditch, M. 1955. Role differentiation in the nuclear family: a comparative study. In *Family, socialization and interaction process,* ed. T. Parsons and R. F. Bales. Glencoe, Ill: Free Press.

Zhukovskaya, R. E. 1976. *Igra i ee pedagogicheskoe znachenie* (The game and its pedagogical significance). Moscow: Academy of Pedagogical Sciences.

Zill, N. 1978. Divorce, marital happiness, and the mental health of children. Findings from the FCD National Survey of Children. Paper prepared for the NIMH Workshop on Divorce and Children. New York: Foundation for Child Development.

Zimbardo, P. G. 1973. On the ethics of intervention and human psychological research: with special reference to the Stanford prison experiment. *Cognition: International Journal of Cognitive Psychology* 2:243–256.

Zimbardo, P. G., Haney, C., Banks, W. C., and Jaffe, D. 1972. Stanford prison experiment. Tape recording. Stanford, Calif.: Philip G. Zimbardo.

Zunich, M. 1961. A study of the relationships between child rearing attitudes and maternal behavior. *Journal of Experimental Education* 30:231–241.

Index